Advances in MATHEMATICAL ECONOMICS

Aims and Scope. The project is to publish *Advances in Mathematical Economics* once a year under the auspices of the Research Center for Mathematical Economics. It is designed to bring together those mathematicians who are seriously interested in obtaining new challenging stimuli from economic theories and those economists who are seeking effective mathematical tools for their research.

The scope of *Advances in Mathematical Economics* includes, but is not limited to, the following fields:

- Economic theories in various fields based on rigorous mathematical reasoning.
- Mathematical methods (e.g., analysis, algebra, geometry, probability) motivated by economic theories.
- Mathematical results of potential relevance to economic theory.
- Historical study of mathematical economics.

Authors are asked to develop their original results as fully as possible and also to give a clear-cut expository overview of the problem under discussion. Consequently, we will also invite articles which might be considered too long for publication in journals.

More information about this series at http://www.springer.com/series/4129

Shigeo Kusuoka • Toru Maruyama
Editors

Advances in Mathematical Economics

Volume 22

Springer

Editors
Shigeo Kusuoka
The University of Tokyo
Tokyo, Japan

Toru Maruyama
Keio University
Tokyo, Japan

ISSN 1866-2226 ISSN 1866-2234 (electronic)
Advances in Mathematical Economics
ISBN 978-981-13-4462-6 ISBN 978-981-13-0605-1 (eBook)
https://doi.org/10.1007/978-981-13-0605-1

Printed on acid-free paper

This Springer imprint is published by the registered company Springer Nature Singapore Pte Ltd.
The registered company address is: 152 Beach Road, #21-01/04 Gateway East, Singapore 189721, Singapore

Contents

Numerical Analysis on Quadratic Hedging Strategies for Normal Inverse Gaussian Models

Takuji Arai, Yuto Imai, and Ryo Nakashima

Abstract The authors aim to develop numerical schemes of the two representative quadratic hedging strategies: locally risk-minimizing and mean-variance hedging strategies, for models whose asset price process is given by the exponential of a normal inverse Gaussian process, using the results of Arai et al. (Int J Theor Appl Financ 19:1650008, 2016) and Arai and Imai (A closed-form representation of mean-variance hedging for additive processes via Malliavin calculus, preprint. Available at https://arxiv.org/abs/1702.07556). Here normal inverse Gaussian process is a framework of Lévy processes that frequently appeared in financial literature. In addition, some numerical results are also introduced.

Keywords Local risk minimization · Mean-variance hedging · Normal inverse Gaussian process · Fast Fourier transform

Article type: Research Article
Received: December 28, 2017
Revised: January 12, 2018

JEL Classification: G11, G12
Mathematics Subject Classification (2010): 91G20, 91G60, 60G51

T. Arai (✉)
Department of Economics, Keio University, Tokyo, Japan
e-mail: arai@econ.keio.ac.jp

Y. Imai
Graduate School of Management, Tokyo Metropolitan University, Tokyo, Japan

R. Nakashima
Power Solutions Inc., Tokyo, Japan

S. Kusuoka, T. Maruyama (eds.), *Advances in Mathematical Economics*, Advances in Mathematical Economics 22, https://doi.org/10.1007/978-981-13-0605-1_1

1 Introduction

Locally risk-minimizing (LRM) and mean-variance hedging (MVH) strategies are well-known quadratic hedging strategies for contingent claims in incomplete markets. In fact, their theoretical aspects have been studied very well for about three decades. On the other hand, numerical methods to compute them have yet to be thoroughly developed. As limited literature, Arai et al. [2] developed a numerical scheme of LRM strategies for call options for two exponential Lévy models: Merton jump-diffusion models and variance gamma (VG) models. Here VG models mean models in which the asset price process is given as the exponential of a VG process. In [2], they made use of a representation for LRM strategies provided by Arai and Suzuki [3] and the so-called Carr-Madan method suggested by [8]: a computational method for option prices using the fast Fourier transforms (FFT). Note that [3] obtained their representation for LRM strategies by means of Malliavin calculus for Lévy processes. As for MVH strategies, Arai and Imai [1] obtained a new closed-form representation for exponential additive models and suggested a numerical scheme for VG models.

Our aim in this paper is to extend the results of [2] and [1] to normal inverse Gaussian (NIG) models. Note that an NIG process is a pure jump Lévy process described as a time-changed Brownian motion as well as a VG process is. Here a process $X = \{X_t\}_{t\geq 0}$ is called a time-changed Brownian motion, if X is described as

$$X_t = \mu Y_t + \sigma B_{Y_t}$$

for any $t \geq 0$, where $\mu \in \mathbb{R}$, $\sigma > 0$, and $B = \{B_t\}_{t\geq 0}$ is a one-dimensional standard Brownian motion and $Y = \{Y_t\}_{t\geq 0}$ is a subordinator, that is, a nondecreasing Lévy process. A time-changed Brownian motion X is called an NIG process, if the corresponding subordinator Y is an inverse Gaussian (IG) process. On the other hand, a VG process is described as a time-changed Brownian motion with Gamma subordinator. NIG process, which has been introduced by Barndorff-Nielsen [4], is frequently appeared in financial literature, e.g., [5–7, 11, 12], and so forth.

Next, we introduce quadratic hedging strategies. Consider a financial market composed of one risk-free asset and one risky asset with finite maturity $T > 0$. For simplicity, we assume that market's interest rate is zero, that is, the price of the risk-free asset is 1 at all times. Let $S = \{S_t\}_{t\in[0,T]}$ be the risky asset price process. Here we prepare some terminologies.

Definition 1.1

1. A strategy is defined as a pair $\varphi = (\xi, \eta)$, where $\xi = \{\xi_t\}_{t\in[0,T]}$ is a predictable process and $\eta = \{\eta_t\}_{t\in[0,T]}$ is an adapted process. Note that ξ_t (resp. η_t) represents the amount of units of the risky asset (resp. the risk-free asset) an investor holds at time t. The wealth of the strategy $\varphi = (\xi, \eta)$ at time $t \in [0, T]$ is given as $V_t(\varphi) := \xi_t S_t + \eta_t$. In particular, $V_0(\varphi)$ gives the initial cost of φ.

2. A strategy φ is said to be self-financing, if it satisfies $V_t(\varphi) = V_0(\varphi) + G_t(\xi)$ for any $t \in [0, T]$, where $G(\xi) = \{G_t(\xi)\}_{t\in[0,T]}$ denotes the gain process induced by ξ, that is, $G_t(\xi) := \int_0^t \xi_u dS_u$ for $t \in [0, T]$. If a strategy φ is self-financing, then η is automatically determined by ξ and the initial cost $V_0(\varphi)$. Thus, a self-financing strategy φ can be described by a pair $(\xi, V_0(\varphi))$.
3. For a strategy φ, a process $C(\varphi) = \{C_t(\varphi)\}_{t\in[0,T]}$ defined by $C_t(\varphi) := V_t(\varphi) - G_t(\xi)$ for $t \in [0, T]$ is called the cost process of φ. When φ is self-financing, its cost process $C(\varphi)$ is a constant.
4. Let F be a square-integrable random variable, which represents the payoff of a contingent claim at the maturity T. A strategy φ is said to replicate claim F, if it satisfies $V_T(\varphi) = F$.

Roughly speaking, a strategy $\varphi^F = (\xi^F, \eta^F)$, which is not necessarily self-financing, is called the LRM strategy for claim F, if it is the replicating strategy minimizing a risk caused by $C(\varphi^F)$ in the L^2-sense among all replicating strategies. Note that it is sufficient to get a representation of ξ^F in order to obtain the LRM strategy φ^F, since η^F is automatically determined by ξ^F. On the other hand, the MVH strategy for claim F is defined as the self-financing strategy minimizing the corresponding L^2-hedging error, that is, the solution (ϑ^F, c^F) to the minimization problem

$$\min_{c,\vartheta} \mathbb{E}\left[(F - c - G_T(\vartheta))^2\right].$$

Remark that c^F gives the initial cost, which is regarded as the corresponding price of F.

In this paper, we propose numerical methods of LRM strategies ξ^F and MVH strategies ϑ^F for call options when the asset price process is given by an exponential NIG process, by extending results of [2] and [1]. Our main contributions are as follows:

1. To ensure the existence of LRM and MVH strategies, we need to impose some integrability conditions (Assumption 1.1 of [2]) with respect to the Lévy measure of the logarithm of the asset price process. Thus, we shall give a sufficient condition in terms of the parameters of NIG processes as our standing assumptions, which enables us to check if a parameter set estimated by financial market data satisfies Assumption 1.1 of [2].
2. The so-called minimal martingale measure (MMM) is indispensable to discuss the LRM problem. In particular, the characteristic function of the asset price process under the MMM is needed in the numerical method developed by [2]. Thus, we provide its explicit representation for NIG models.
3. In general, a Fourier transform is given as an integration on $[0, \infty)$. In fact, we represent LRM strategies by such an improper integration and truncate its integration interval in order to use FFTs. Thus, we shall estimate a sufficient length of the integration interval to reduce the associated truncation error within given allowable extent.

Actually, we need to overcome some complicated calculations in order to achieve the three objects above, since the Lévy measure of an NIG process includes a modified Bessel function of the second kind with parameter 1.

An outline of this paper is as follows: A precise model description is given in Sect. 2. Main results will be stated in Sect. 3. Our standing assumption described in terms of the parameters of NIG models is introduced in Sect. 3.1, which is followed by subsections discussing the characteristic function under the MMM, a representation of LRM strategies, an estimation of the integration interval, and a representation of MVH strategies. Note that proofs are postponed until Appendix. Sect. 4 is devoted to numerical results.

2 Model Description

We consider throughout a financial market composed of one risk-free asset and one risky asset with finite time horizon $T > 0$. For simplicity, we assume that market's interest rate is zero, that is, the price of the risk-free asset is 1 at all times. $(\Omega, \mathscr{F}, \mathbb{P})$ denotes the canonical Lévy space, which is given as the product space of spaces of compound Poisson processes on $[0, T]$. Denote by $\mathbb{F} = \{\mathscr{F}_t\}_{t\in[0,T]}$ the canonical filtration completed for $\mathbb{P}$. For more details on the canonical Lévy space, see Section 4 of Solé et al. [16] or Section 3 of Delong and Imkeller [10]. Let $L = \{L_t\}_{t\in[0,T]}$ be a pure jump Lévy process with Lévy measure ν defined on $(\Omega, \mathscr{F}, \mathbb{P})$. We define the jump measure of L as

$$N([0,t], A) := \sum_{0\leq u\leq t} \mathbf{1}_A(\Delta L_u)$$

for any $A \in \mathscr{B}(\mathbb{R}_0)$ and any $t \in [0, T]$, where $\Delta L_t := L_t - L_{t-}$, $\mathbb{R}_0 := \mathbb{R} \setminus \{0\}$, and $\mathscr{B}(\mathbb{R}_0)$ denotes the Borel σ-algebra on $\mathbb{R}_0$. In addition, its compensated version $\widetilde{N}$ is defined as

$$\widetilde{N}([0,t], A) := N([0,t], A) - t\nu(A).$$

In this paper, we study the case where L is given as a normal inverse Gaussian (NIG) process. Here a pure jump Lévy process L is called an NIG process with parameters $\alpha > 0$, $-\alpha < \beta < \alpha$, and $\delta > 0$, if its characteristic function is given as

$$\mathbb{E}[e^{izL_t}] = \exp\left\{-\delta\left(\sqrt{\alpha^2 - (\beta + iz)^2} - \sqrt{\alpha^2 - \beta^2}\right)\right\}$$

for any $z \in \mathbb{C}$ and any $t \in [0, T]$. Note that the corresponding Lévy measure ν is given as

$$\nu(dx) = \frac{\delta\alpha}{\pi}\frac{e^{\beta x}K_1(\alpha|x|)}{|x|}dx$$

for $x \in \mathbb{R}_0$, where K_1 is the modified Bessel function of the second kind with parameter 1. When we need to emphasize the model parameters, ν is denoted by $\nu[\alpha, \beta, \delta]$. In addition, the process L can also be described as the following time-changed Brownian motion with IG subordinator:

$$L_t = \beta\delta^2 I_t + \delta B_{I_t},$$

where $B = \{B_t\}_{t\in[0,T]}$ is a one-dimensional standard Brownian motion and $I = \{I_t\}_{t\in[0,T]}$ is an IG process with parameter $(1, \delta\sqrt{\alpha^2 - \beta^2})$. For more details on NIG processes, see Section 4.4 of Cont and Tankov [9] and Subsection 5.3.8 of Schoutens [13]. In this paper, the risky asset price process $S = \{S_t\}_{t\in[0,T]}$ is given as the exponential of the NIG process L:

$$S_t = S_0 e^{L_t},$$

where $S_0 > 0$.

Now, we prepare some additional notation. For $v \in [0, \infty)$ and $a \in (\frac{3}{2}, 2]$, we define

$$M_1(v, a) := \frac{v^2 + \alpha^2 - (a + \beta)^2}{\alpha^2}, \quad M_2 := 1 - \frac{\beta^2}{\alpha^2}, \quad b(v, a) := \frac{2(a + \beta)v}{\alpha^2},$$

and

$$W(v, a) := \frac{\delta\alpha}{\sqrt{2}}\left(i\sqrt{\sqrt{M_1^2 + b^2} - M_1} - \sqrt{\sqrt{M_1^2 + b^2} + M_1} + \sqrt{2M_2}\right), \tag{1}$$

where $M_1(v, a)$ and $b(v, a)$ are abbreviated to M_1 and b, respectively. Note that we can define $W(0, 1)$ and $W(v, a + 1)$ for $v \in [0, \infty)$ and $a \in (\frac{3}{2}, 2]$ as well. Moreover, when it is desirable to emphasize the parameters α, β, and δ, we denote the above four functions as $M_1(v, a; \alpha, \beta)$, $M_2(\alpha, \beta)$, $b(v, a; \alpha, \beta)$, and $W(v, a; \alpha, \beta, \delta)$, respectively.

3 Main Results

3.1 Standing Assumption

We introduce our standing assumption in terms of model parameters.

Assumption 3.1

$$\alpha > \frac{5}{2}, \quad -\frac{3}{2} < \beta \leq -\frac{1}{2}, \quad \text{and} \quad \beta + 4 < \alpha.$$

Now, we show that Assumption 3.1 is a sufficient condition for Assumption 1.1 of [2], which ensures the existence of LRM and MVH strategies.

Proposition 3.1 *Under Assumption* 3.1, *we have*

1. $\int_{\mathbb{R}_0}(e^x-1)^4\nu(dx)<\infty$,
2. $0\geq\int_{\mathbb{R}_0}(e^x-1)\nu(dx)>-\int_{\mathbb{R}_0}(e^x-1)^2\nu(dx)$.

We postpone the proof of Proposition 3.1 until Appendix. Remark that Condition 2 in Proposition 3.1 is the same as the second condition of Assumption 1.1 of [2]. On the other hand, Condition 1 is a modification of the first condition of Assumption 1.1 of [2], which is given as follows:

$$\int_{\mathbb{R}_0}(|x|\vee x^2)\nu(dx)<\infty \text{ and } \int_{\mathbb{R}_0}(e^x-1)^n\nu(dx)<\infty \text{ for } n=2,4.$$

Firstly, $\int_{\mathbb{R}_0}x^2\nu(dx)<\infty$ and $\int_{\mathbb{R}_0}(e^x-1)^2\nu(dx)<\infty$ are redundant, since $\int_{\mathbb{R}_0}(x^2\wedge 1)\nu(dx)<\infty$ holds. Next, NIG processes do not have the finiteness of $\int_{\mathbb{R}_0}|x|\nu(dx)$, different from VG processes. Actually, S is described by a stochastic integration with respect to N in [2]. Thus, the condition $\int_{\mathbb{R}_0}|x|\nu(dx)<\infty$ is needed. On the other hand, describing S as

$$S_t=S_0e^{L_t}=S_0\exp\left\{\int_0^t\int_{\mathbb{R}_0}x\widetilde{N}(du,dx)+t\int_{\mathbb{R}_0}x\nu(dx)\right\},$$

we do not need to assume it.

3.2 The Minimal Martingale Measure

In this subsection, we focus on the minimal martingale measure (MMM): an equivalent martingale measure under which any square-integrable $\mathbb{P}$-martingale orthogonal to the martingale part of S remains a martingale. Remark that the MMM plays a vital role in quadratic hedging problems. Denote $\mu^S:=\int_{\mathbb{R}_0}(e^x-1)\nu(dx)$, $C_\nu:=\int_{\mathbb{R}_0}(e^x-1)^2\nu(dx)$, $h:=\mu^S/C_\nu$, and

$$\theta_x:=\frac{\mu^S(e^x-1)}{C_\nu}$$

for $x\in\mathbb{R}_0$. As discussed in [2], the MMM $\mathbb{P}^*$ exists under Assumption 1.1 of [2], and its Radon-Nikodym density is given as

$$\frac{d\mathbb{P}^*}{d\mathbb{P}}=\exp\left\{\int_{\mathbb{R}_0}\log(1-\theta_x)\widetilde{N}([0,T],dx)+T\int_{\mathbb{R}_0}\left(\log(1-\theta_x)+\theta_x\right)\nu(dx)\right\}.$$

Note that $\theta_x < 1$ holds for any $x \in \mathbb{R}_0$ under Assumption 3.1 by Proposition 3.1. Furthermore, $\mathbb{P}^*$ is not only the MMM but also the variance-optimal martingale measure (VOMM) in our setting as discussed in [1]. Note that the VOMM is an equivalent martingale measure whose density minimizes the $L^2(\mathbb{P})$-norm among all equivalent martingale measures. Since MVH strategies are described using the VOMM, we use $\mathbb{P}^*$ to express MVH strategies as well as LRM strategies.

Here we prepare some additional notation. From the view of the Girsanov theorem,

$$\widetilde{N}^{\mathbb{P}^*}([0,t],dx) := \widetilde{N}([0,t],dx) + \theta_x \nu(dx)t$$

is the compensated jump measure of L under $\mathbb{P}^*$. This means that the Lévy measure under $\mathbb{P}^*$, denoted by $\nu^{\mathbb{P}^*}$, is given as

$$\nu^{\mathbb{P}^*}(dx) = (1-\theta_x)\nu(dx). \tag{2}$$

L is then rewritten as

$$L_t = \int_{\mathbb{R}_0} x\widetilde{N}^{\mathbb{P}^*}([0,t],dx) + \mu^* t, \tag{3}$$

where $\mu^* := \int_{\mathbb{R}_0}(x - e^x + 1)\nu^{\mathbb{P}^*}(dx)$ and the stochastic differential equation for S under $\mathbb{P}^*$ is given as $dS_t = S_{t-}\int_{\mathbb{R}_0}(e^x - 1)\widetilde{N}^{\mathbb{P}^*}(dt,dx)$.

In order to develop FFT-based numerical schemes, we need an explicit representation of the characteristic function of L under $\mathbb{P}^*$:

$$\phi_{T-t}(z) := \mathbb{E}_{\mathbb{P}^*}[e^{izL_{T-t}}]$$

for $z \in \mathbb{C}$. Before stating it, we calculate $\nu^{\mathbb{P}^*}(dx)$ the Lévy measure of L under $\mathbb{P}^*$. Recall that $\nu[\alpha,\beta,(1+h)\delta](dx)$ represents the Lévy measure of an NIG process with parameters α, β, and $(1+h)\delta$. We provide the proof of the following proposition in Appendix.

Proposition 3.2 *We have*

$$\nu^{\mathbb{P}^*}(dx) = \nu[\alpha,\beta,(1+h)\delta](dx) + \nu[\alpha,1+\beta,-h\delta](dx).$$

Now, we provide a representation of ϕ using the function $W(v,a)$ defined in (1). Remark that $W(v,a;\alpha,1+\beta,\delta)$ is also well-defined, since $M_2(\alpha,\beta+1) > 0$ by Assumption 3.1. The proof of the following proposition is given in Appendix.

Proposition 3.3 *For any* $v \in [0, \infty)$ *and any* $a \in (\frac{3}{2}, 2]$, *we have*

$$\phi_{T-t}(v-ia) = \exp\left\{(T-t)i(v-ia)\left(\mu^* - \frac{(1+h)\delta\beta}{\sqrt{\alpha^2-\beta^2}} + \frac{h\delta(1+\beta)}{\sqrt{\alpha^2-(1+\beta)^2}}\right)\right\}$$
$$\times \exp\left\{(T-t)\Big(W(v,a;\alpha,\beta,(1+h)\delta) + W(v,a;\alpha,1+\beta,-h\delta)\Big)\right\}$$

where $\mu^* = \int_{\mathbb{R}_0}(x - e^x + 1)\nu^{\mathbb{P}^*}(dx)$.

3.3 Local Risk Minimization

In this subsection, we introduce how to compute LRM strategies for call options $(S_T - K)^+$ with strike price $K > 0$. First of all, we give a precise definition of the LRM strategy for claim $F \in L^2(\mathbb{P})$. The following is based on Theorem 1.6 of Schweizer [15].

Definition 3.1

1. A strategy $\varphi = (\xi, \eta)$ is said to be an L^2-strategy, if ξ satisfies $\mathbb{E}\left[\int_0^T S_{u-}^2\xi_u^2 du\right] < \infty$ and $V(\varphi)$ is a right continuous process with $\mathbb{E}[V_t^2(\varphi)] < \infty$ for every $t \in [0, T]$.
2. An L^2-strategy φ is called the LRM strategy for claim F, if $V_T(\varphi^F) = F$ and $[C(\varphi^F), M]$ is a uniformly integrable martingale, where $M = \{M_t\}_{t\in[0,T]}$ is the martingale part of S.

Note that all the conditions of Theorem 1.6 of [15] hold under Assumption 1.1 of [2] as seen in Example 2.8 of [3]. The above definition of LRM strategies is a simplified version, since the original one, introduced in [14] and [15], is rather complicated. Now, an $F \in L^2(\mathbb{P})$ admits a Föllmer-Schweizer decomposition, if it can be described by

$$F = F_0 + G_T(\xi^{FS}) + L_T^{FS},$$

where $F_0 \in \mathbb{R}$, $\xi^{FS} = \{\xi_t^{FS}\}_{t\in[0,T]}$ is a predictable process satisfying $\mathbb{E}\left[\int_0^T S_{u-}^2(\xi_u^{FS})^2 du\right] < \infty$, and $L^{FS} = \{L_t^{FS}\}_{t\in[0,T]}$ is a square-integrable martingale orthogonal to M with $L_0^{FS} = 0$. In addition, Proposition 5.2 of [15] provides that, under Assumption 1.1 of [2], the LRM strategy $\varphi^F = (\xi^F, \eta^F)$ for $F \in L^2(\mathbb{P})$ exists if and only if F admits a Föllmer-Schweizer decomposition; and its relationship is given by

$$\xi_t^F = \xi_t^{FS}, \quad \eta_t^F = F_0 + G_t(\xi^F) + L_t^{FS} - \xi_t^F S_t.$$

As a result, it suffices to obtain a representation of ξ^F in order to get φ^F. Henceforth, we identify ξ^F with φ^F.

We consider call options $(S_T - K)^+$ with strike price $K > 0$ as claims to hedge. Now, we denote $F(K) = (S_T - K)^+$ for $K > 0$ and define a function

$$I(s,t,K) := \int_{\mathbb{R}_0} \mathbb{E}_{\mathbb{P}^*}[(S_T e^x - K)^+ - (S_T - K)^+ | S_{t-} = s](e^x - 1)\nu(dx)$$

for $s > 0$, $t \in [0, T]$, and $K > 0$. [3] gave an explicit representation of $\xi_t^{F(K)}$ for any $t \in [0, T]$ and any $K > 0$ using Malliavin calculus for Lévy processes.

Proposition 3.4 (Proposition 4.6 of [3]) *For any $K > 0$ and any $t \in [0, T]$,*

$$\xi_t^{F(K)} = \frac{I(S_{t-}, t, K)}{S_{t-} C_\nu}. \tag{4}$$

In addition, [2] introduced an integral representation for $I(S_{t-}, t, K)$ as

$$I(S_{t-},t,K) = \frac{1}{\pi}\int_0^\infty K^{-iv-a+1}\int_{\mathbb{R}_0}(e^{(iv+a)x}-1)(e^x-1)\nu(dx)\frac{\phi_{T-t}(v-ia)S_{t-}^{iv+a}}{(iv+a)(iv+a-1)}dv,$$

where $a \in (1, 2]$ and the right-hand side is independent of the choice of a. Remark that we narrow the range of a to $(\frac{3}{2}, 2]$ for technical reasons, but this does not restrict our development of numerical schemes, since we take 1.75 as the value of a in our numerical experiments. To compute $I(S_{t-}, t, K)$, we need to calculate the integration $\int_{\mathbb{R}_0}(e^{(iv+a)x} - 1)(e^x - 1)\nu(dx)$. Now, Lemma A.1 implies that

$$\begin{aligned}
\int_{\mathbb{R}_0} e^{(iv+a)x}(e^x - 1)\nu(dx) &= \int_{\mathbb{R}_0} e^{(iv+a)x}(e^x - 1)\nu(dx) \\
&= \int_{\mathbb{R}_0} (e^{(iv+a+1)x} - e^{(iv+a)x})\nu(dx) \\
&= \int_{\mathbb{R}_0} (e^{(iv+a+1)x} - 1)\nu(dx) \\
&\quad - \int_{\mathbb{R}_0} (e^{(iv+a)x} - 1)\nu(dx) \\
&= W(v, a+1) - W(v, a),
\end{aligned}$$

from which we have

$$\begin{aligned}
I(S_{t-}, t, K) = \frac{1}{\pi}\int_0^\infty & K^{-iv-a+1}\Big(W(v, a+1) - W(v, a) - W(0, 1)\Big) \\
&\times \frac{\phi_{T-t}(v - ia)S_{t-}^{iv+a}}{(iv+a)(iv+a-1)}dv.
\end{aligned} \tag{5}$$

Thus, we can compute $I(S_{t-}, t, K)$ using the FFT as mentioned in [2].

3.4 Integration Interval

To compute the integral (5) with the FFT, we discretize $I(S_{t-}, t, K)$ as

$$I(S_{t-}, t, K) \approx \frac{1}{\pi} \sum_{j=0}^{N-1} e^{(-i\eta j - a+1)\log K} \Big(W(\eta j, a+1) - W(\eta j, a) - W(0,1) \Big) \times \frac{\phi_{T-t}(\eta j - ia) S_{t-}^{i\eta j + a}}{(i\eta j + a)(i\eta j + a - 1)} \eta,$$

where N represents the number of grid points and $\eta > 0$ is the distance between adjacent grid points. This approximation corresponds to the integral (5) over the interval $[0, N\eta]$, so we need to specify N and η to satisfy

$$\left| \frac{1}{\pi} \int_{N\eta}^{\infty} K^{-iv-a+1} \Big(W(v, a+1) - W(v, a) - W(0,1) \Big) \frac{\phi_{T-t}(v - ia) S_{t-}^{iv+a}}{(iv + a)(iv + a - 1)} dv \right| < \varepsilon \tag{6}$$

for a given sufficiently small value $\varepsilon > 0$, which represents the allowable error. Thus, we shall estimate a sufficient length for the integration interval of (5) for a given allowable error $\varepsilon > 0$ in the sense of (6). The following proposition is shown in Appendix.

Proposition 3.5 *For $\varepsilon > 0$ and $t \in [0, T)$, if $w > 1$ satisfies*

$$\frac{\sqrt{2} K^{-a+1} S_{t-}^{a} C(t)}{\pi (T - t)\varepsilon} \left(2 + \sqrt{\alpha^2 - (a+\beta)^2 + 2(a+1+\beta)^2} \right) < e^{(T-t)\delta w}, \tag{7}$$

we have

$$\left| \frac{1}{\pi} \int_{w}^{\infty} K^{-iv-a+1} \Big(W(v, a+1) - W(v, a) - W(0,1) \Big) \frac{\phi_{T-t}(v - ia) S_{t-}^{iv+a}}{(iv + a)(iv + a - 1)} dv \right| < \varepsilon,$$

where $C(t)$ is defined as

$$C(t) := \exp\left\{ (T-t)a \left(\mu^* - \frac{(1+h)\delta\beta}{\sqrt{\alpha^2 - \beta^2}} + \frac{h\delta(1+\beta)}{\sqrt{\alpha^2 - (1+\beta)^2}} \right) \right\} \times \exp\left\{ (T-t)\delta\alpha \left((1+h)\sqrt{M_2(\alpha, \beta)} - h\sqrt{M_2(\alpha, 1+\beta)} \right) \right\} \tag{8}$$

for any $t \in [0, T)$.

Remark 3.1 In Proposition 3.5, the case of $t = T$ is excluded, but this does not restrict our numerical method, since we do not need to compute the value of LRM strategies when the time to maturity $T - t$ is 0.

3.5 Mean-Variance Hedging

As introduced in Sect. 1, the MVH strategy for claim $F \in L^2(\mathbb{P})$ is defined as the solution (ϑ^F, c^F) to the minimization problem

$$\min_{c\in\mathbb{R},\vartheta\in\Theta} \mathbb{E}\left[\left(F - c - G_T(\vartheta)\right)^2\right],$$

where Θ is the set of all admissible strategies, mathematically the set of $\mathbb{R}$-valued S-integrable predictable processes ϑ satisfying $\mathbb{E}\left[\int_0^T \vartheta_u^2 S_{u-}^2 du\right] < \infty$. Arai and Imai [1] gave an explicit closed-form representation of ϑ^F for exponential additive models and developed a numerical scheme for call options $(S_T - K)^+$ with strike price $K > 0$ for exponential Lévy models. Different from LRM strategies, the value of ϑ_t^F is depending on not only S_{t-} but also the whole trajectory of S from 0 to $t-$. However it is impossible to observe the trajectory of S continuously. Thus, [1] developed a numerical scheme to compute ϑ_t^F approximately using discrete observational data $S_{t_0}, S_{t_1}, \ldots, S_{t_n}$, where $n \geq 1$ and $t_k := \frac{kt}{n+1}$.

We need some preparations before introducing the representation of ϑ_t^F obtained by [1]. Firstly, we consider the VOMM, which is an equivalent martingale measure whose density minimizes the $L^2(\mathbb{P})$-norm among all equivalent martingale measures. Indeed, the MMM $\mathbb{P}^*$ coincides with the VOMM in our setting as mentioned in Sect. 3.2. Next, we define a process $\mathscr{E} = \{\mathscr{E}_t\}_{t\in[0,T]}$ as a solution to the stochastic differential equation $\mathscr{E}_t = 1 - h\int_0^t \mathscr{E}_{u-} dS_u$, and $H^F = \{H_t^F\}_{t\in[0,T]}$ as $H_t^F := \mathbb{E}_{\mathbb{P}^*}[F|S_{t-}]$. Moreover, remark that Assumption 2.1 of [1] is satisfied under Assumption 3.1.

From the view of [1], the MVH strategy $\vartheta_t^{F(K)}$ for call option $F(K) = (S_T - K)^+$ is represented in closed-form as

$$\vartheta_t^{F(K)} = \xi_t^{F(K)} + \frac{h\mathscr{E}_{t-}}{S_{t-}} \int_0^{t-} \frac{dH_u^{F(K)} - \xi_u^{F(K)} dS_u}{\mathscr{E}_u}.$$

Now, the process $H_t^{F(K)} = \mathbb{E}_{\mathbb{P}^*}[F(K)|S_{t-}]$ is represented as

$$H_t^{F(K)} = \frac{1}{\pi}\int_0^\infty K^{-iv-a+1} \frac{\phi_{T-t}(v - ia) S_{t-}^{iv+a}}{(iv + a - 1)(iv + a)} dv,$$

which is computable with the FFT. As a result, using discrete observational data $S_{t_0}, S_{t_1}, \ldots, S_{t_n}$, we can approximate $\vartheta_t^{F(K)}$ as

$$\vartheta_t^{F(K)} \approx \xi_t^{F(K)} + \frac{h\mathscr{E}_{t_n}}{S_{t_n}} \sum_{k=1}^n \frac{\Delta H_{t_k}^{F(K)} - \xi_{t_k}^{F(K)} \Delta S_{t_k}}{\mathscr{E}_{t_k}}, \tag{9}$$

where $H_{t_k}^{F(K)} = \mathbb{E}_{\mathbb{P}^*}[F(K)|S_{t_k}]$ and $t_k := \frac{kt}{n+1}$ for $k = 0, 1, \ldots, n$; t is corresponding to t_{n+1}; and, for $k = 1, \ldots, n$, we denote $\Delta X_{t_k} := X_{t_k} - X_{t_{k-1}}$ for a process X and

$$\mathscr{E}_{t_{k+1}} = \mathscr{E}_{t_k}\left\{1 - \frac{h\Delta S_{t_{k+1}}}{S_{t_k}}\right\}$$

with $\mathscr{E}_{t_0} = 1$.

4 Numerical Results

We consider European call options on the S&P 500 Index (SPX) matured on 19 May 2017 and set the initial date of our hedging to 20 May 2016. We fix T to 1. There are 250 business days on and after 20 May 2016 until and including 19 May 2017. For example, 20 May 2016 and 23 May 2016 are corresponding to time 0 and $\frac{1}{249}$, respectively, since 20 May 2016 is Friday. Note that we shall use 250 dairy closing prices of the SPX on and after 20 May 2016 until and including 19 May 2017 as discrete observational data. Figure 1 illustrates the fluctuation of the SPX.

Next, we set model parameters as

$$\begin{cases} \alpha = 25.61598030765035, \\ \beta = -1.2668546614155765, \\ \delta = 0.40532772478162127, \end{cases}$$

which are calibrated by the data set of European call options on the SPX at 20 April 2016. Note that the above parameter set satisfies Assumption 3.1. Moreover, we choose

$$N = 2^{16}, \eta = 0.25, \text{ and } a = 1.75$$

as parameters related to the FFT, that is, $N\eta = 2^{14}$, which satisfies (7) for any $t \leq \frac{248}{249}$ when we take $\varepsilon = 0.01$ as our allowable error.

As contingent claims to hedge, we consider call options with strike price K =2300, 2350, and 2400 and compute the values of LRM strategies $\xi_t^{F(K)}$ and MVH strategies $\vartheta_t^{F(K)}$ for $t = \frac{1}{249}, \frac{2}{249}, \ldots, 1$ by using (4), (5), and (9). Remark that, for $k = 1, \ldots, 249$, $\xi_{\frac{k}{249}}^{F(K)}$ and $\vartheta_{\frac{k}{249}}^{F(K)}$ are constructed on time $\frac{k-1}{249}$ using observational data $S_0, S_{\frac{1}{249}}, \ldots, S_{\frac{k-1}{249}}$. Figures 2, 3, and 4 show the values of $\xi_t^{F(K)}$ and $\vartheta_t^{F(K)}$ versus times $t = \frac{1}{249}, \frac{2}{249}, \ldots, 1$ for the case where $K = 2300, 2350$, and 2400, respectively.

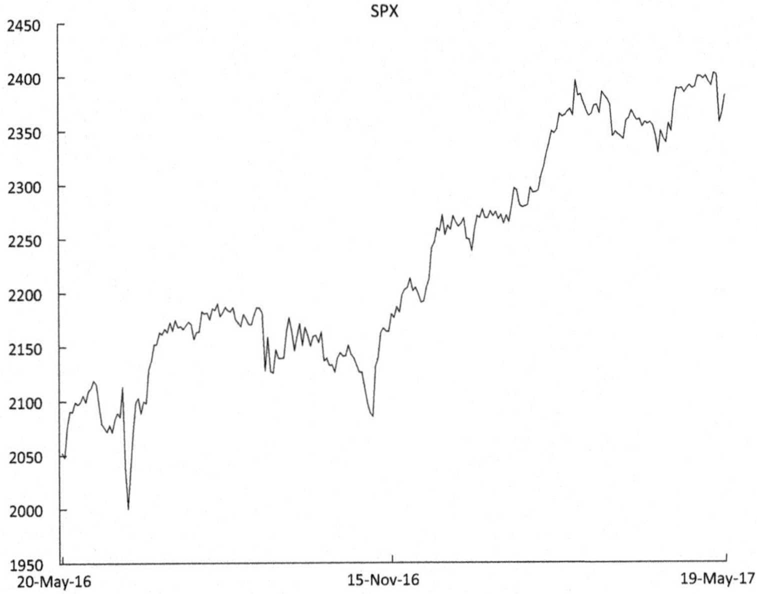

Fig. 1 SPX dairy closing prices

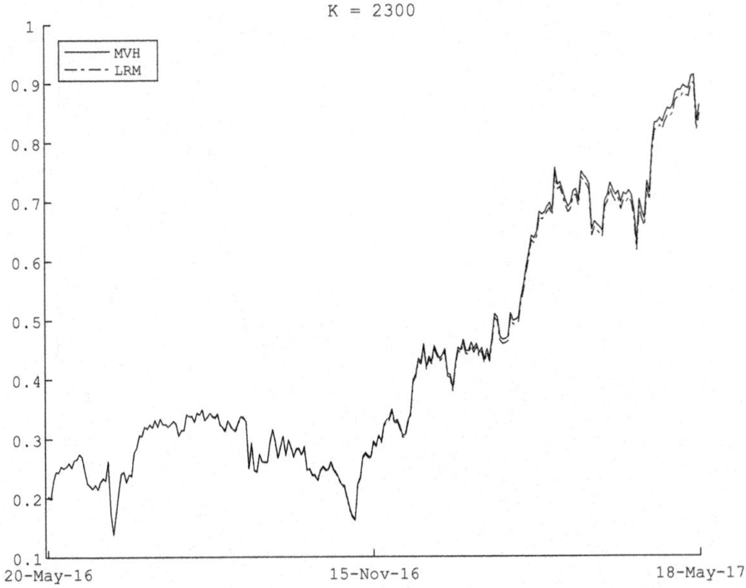

Fig. 2 Values of LRM strategies $\xi_t^{F(K)}$ and MVH strategies $\vartheta_t^{F(K)}$ for $K = 2300$. The dotted and the solid lines represent the values of $\xi_t^{F(K)}$ and $\vartheta_t^{F(K)}$, respectively. The two lines are almost overlapping when t is small and separate gradually as drawing near to the maturity

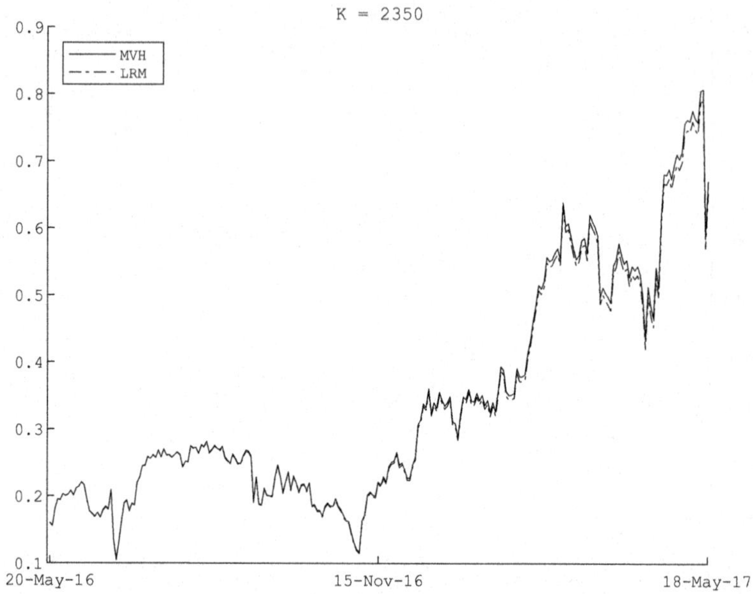

Fig. 3 Values of LRM strategies $\xi_t^{F(K)}$ and MVH strategies $\vartheta_t^{F(K)}$ for $K = 2350$

Fig. 4 Values of LRM strategies $\xi_t^{F(K)}$ and MVH strategies $\vartheta_t^{F(K)}$ for $K = 2400$

Appendix

Proof of Proposition 3.1

In order to see Condition 1, it suffices to show $\int_1^\infty (e^x - 1)^4 \nu(dx) < \infty$ and $\int_{-\infty}^{-1} (e^x - 1)^4 \nu(dx) < \infty$.

Firstly, we see $\int_1^\infty (e^x - 1)^4 \nu(dx) < \infty$. Noting that the Sommerfeld integral representation for the function K_1 (see, e.g., Appendix A of [9]):

$$K_1(z) = \frac{z}{4} \int_0^\infty \exp\left\{-s - \frac{z^2}{4s}\right\} s^{-2} ds \tag{10}$$

for $z \geq 0$, we have

$$\begin{aligned}
\int_1^\infty (e^x - 1)^4 \nu(dx) &= \frac{\delta\alpha}{\pi} \int_1^\infty (e^x - 1)^4 \frac{e^{\beta x} K_1(\alpha x)}{x} dx \\
&= \frac{\delta\alpha}{\pi} \int_\alpha^\infty (e^{\frac{z}{\alpha}} - 1)^4 \exp\left\{\frac{\beta}{\alpha} z\right\} \frac{1}{z} \int_0^\infty \frac{z}{4} \exp\left\{-s - \frac{z^2}{4s}\right\} s^{-2} ds dz \\
&\leq \frac{\delta\alpha}{4\pi} \int_\alpha^\infty \exp\left\{\frac{4+\beta}{\alpha} z\right\} \int_0^\infty \frac{z}{\alpha} \exp\left\{-s - \frac{z^2}{4s}\right\} s^{-2} ds dz \\
&= \frac{\delta}{4\pi} \int_0^\infty e^{-s} s^{-2} \int_\alpha^\infty \frac{z}{\sqrt{2\pi 2s}} \exp\left\{-\frac{1}{4s}\left(z - 2s\frac{4+\beta}{\alpha}\right)^2\right\} dz \\
&\quad \times \exp\left\{\left(\frac{4+\beta}{\alpha}\right)^2 s\right\} \sqrt{2\pi 2s} ds \\
&\leq \frac{\delta}{4\pi} \int_0^\infty e^{-s} s^{-2} \cdot 2s \frac{4+\beta}{\alpha} \cdot \exp\left\{\left(\frac{4+\beta}{\alpha}\right)^2 s\right\} \sqrt{2\pi 2s} dt \\
&= \frac{\delta}{\sqrt{\pi}} \frac{4+\beta}{\alpha} \int_0^\infty s^{-\frac{1}{2}} \exp\left\{\left(\left(\frac{4+\beta}{\alpha}\right)^2 - 1\right) s\right\} ds \\
&= \delta(4+\beta)\left(\alpha^2 - (4+\beta)^2\right)^{-\frac{1}{2}} < \infty.
\end{aligned}$$

Remark that the above first inequality is given from $(e^{\frac{z}{\alpha}} - 1)^4 \leq e^{\frac{4z}{\alpha}}$ for any $z \in [\alpha, \infty)$.

Next, we show $\int_{-\infty}^{-1} (e^x - 1)^4 \nu(dx) < \infty$ by a similar argument to the above. Noting that $(e^{\frac{z}{\alpha}} - 1)^4 \leq 1$ for any $z \in (-\infty, -\alpha]$, we have

$$\int_{-\infty}^{-1}(e^x-1)^4\nu(dx)\leq\frac{\delta\alpha}{4\pi}\int_{-\infty}^{-\alpha}(e^{\frac{z}{\alpha}}-1)^4\exp\left\{\frac{\beta}{\alpha}z\right\}\int_0^\infty\frac{z}{\alpha}\exp\left\{-s-\frac{z^2}{4s}\right\}s^{-2}dsdz$$
$$\leq\frac{\delta}{4\pi}\int_0^\infty e^{-s}s^{-2}\int_{-\infty}^\infty\frac{z}{\sqrt{2\pi 2s}}\exp\left\{-\frac{1}{4s}\left(z-2s\frac{\beta}{\alpha}\right)^2\right\}dz$$
$$\times\exp\left\{\frac{\beta^2}{\alpha^2}s\right\}\sqrt{2\pi 2s}ds<\infty.$$

Thus, Condition 1 holds true.

To confirm Condition 2, we need some preparations. The following lemma is proven later.

Lemma A.1 *For any $v\in[0,\infty)$ and any $a\in(\frac{3}{2},2]$, we have*

$$\int_{\mathbb{R}_0}\left(e^{(iv+a)x}-1\right)\nu(dx)=W(v,a). \tag{11}$$

In addition, (11) *still holds for the case where* $(v,a)=(0,1)$ *and* $(v,a+1)$.

We have $\int_{\mathbb{R}_0}(e^x-1)\nu(dx)=W(0,1)=\delta\alpha(\sqrt{M_2}-\sqrt{M_1(0,1)})$. Assumption 3.1 implies

$$M_2-M_1(0,1)=\frac{1}{\alpha^2}\left((\alpha^2-\beta^2)-(\alpha^2-(1+\beta)^2)\right)=\frac{1+2\beta}{\alpha^2}\leq 0,$$

from which the inequality $0\geq\int_{\mathbb{R}_0}(e^x-1)\nu(dx)$ holds true. To see the second inequality, since we have

$$-\int_{\mathbb{R}_0}(e^x-1)^2\nu(dx)=-\int_{\mathbb{R}_0}\left((e^{2x}-1)-2(e^x-1)\right)\nu(dx)=-W(0,2)+2W(0,1),$$

it suffices to show $W(0,2)-W(0,1)>0$. Firstly, we have

$$W(0,2)-W(0,1)=\frac{\delta\alpha}{\sqrt{2}}\left(\left(-\sqrt{2M_1(0,2)}+\sqrt{2M_2}\right)-\left(-\sqrt{2M_1(0,1)}+\sqrt{2M_2}\right)\right)$$
$$=\delta\alpha\left(-\sqrt{M_1(0,2)}+\sqrt{M_1(0,1)}\right).$$

On the other hand, it holds that

$$M_1(0,1)-M_1(0,2)=\frac{\alpha^2-(1+\beta)^2-\alpha^2+(2+\beta)^2}{\alpha^2}=\frac{3+2\beta}{\alpha^2}>0$$

by Assumption 3.1. As a result, the inequality $\int_{\mathbb{R}_0}(e^x-1)\nu(dx)>-\int_{\mathbb{R}_0}(e^x-1)^2\nu(dx)$ holds under Assumption 3.1. □

Proof of Lemma A.1

We begin with the following lemma:

Lemma A.2 *For any $\gamma \geq 0$ and any $M > 0$, we have*

$$\int_0^\gamma \int_0^\infty e^{(iu-M)s} s^{-\frac{1}{2}} ds du = \sqrt{2\pi}\left(\sqrt{\sqrt{M^2+\gamma^2}-M}\right.$$
$$\left.+ i\left(\sqrt{\sqrt{M^2+\gamma^2}+M} - \sqrt{2M}\right)\right).$$

Proof Remark that the characteristic function of the Gamma distribution with parameters $\theta > 0$ and $k > 0$ is given as

$$\int_0^\infty e^{iux} \frac{\theta^k}{\Gamma(k)} x^{k-1} e^{-\theta x} dx = \left(\frac{\theta}{\theta - iu}\right)^k$$

for any $u \in \mathbb{R}$, where $\Gamma(\cdot)$ is the Gamma function. We have then

$$\int_0^\infty e^{(iu-M)s} s^{-\frac{1}{2}} ds = \sqrt{\frac{M}{M-iu}} \frac{\Gamma\left(\frac{1}{2}\right)}{\sqrt{M}} = \frac{\sqrt{\pi}}{\sqrt{M-iu}}$$

for any $M > 0$ and any $u \in \mathbb{R}$. Thus, we obtain

$$\int_0^\infty e^{(iu-M)s} s^{-\frac{1}{2}} dt = \sqrt{\frac{\pi}{2}}\left(\frac{\sqrt{\sqrt{M^2+u^2}+M}}{\sqrt{M^2+u^2}} + i\frac{\sqrt{\sqrt{M^2+u^2}-M}}{\sqrt{M^2+u^2}}\right).$$

Putting $x = \sqrt{M^2+u^2}$, we have

$$\int_0^\gamma \frac{\sqrt{\sqrt{M^2+u^2}+M}}{\sqrt{M^2+u^2}} du = \int_M^{\sqrt{M^2+\gamma^2}} \frac{\sqrt{x+M}}{\sqrt{x^2-M^2}} dx = 2\sqrt{\sqrt{M^2+\gamma^2}-M}$$

and

$$\int_0^\gamma \frac{\sqrt{\sqrt{M^2+u^2}-M}}{\sqrt{M^2+u^2}} du = 2\sqrt{\sqrt{M^2+\gamma^2}+M} - 2\sqrt{2M}.$$

This completes the proof of Lemma A.2. □

Now, let us go back to the proof of Lemma A.1. For any $v \in [0, \infty)$ and any $a \in (\frac{3}{2}, 2]$, the same sort of argument as in the proof of Proposition 3.1 implies that

$$\int_{\mathbb{R}_0}\left(e^{(iv+a)x}-1\right)\nu(dx)=\frac{\delta\alpha}{2\sqrt{\pi}}\int_0^\infty e^{-s}s^{-\frac{3}{2}}\int_{\mathbb{R}_0}\frac{e^{(iv+a)\frac{z}{\alpha}}-1}{\sqrt{2\pi 2s}}$$
$$\times\exp\left\{-\frac{1}{4s}\left(z-2s\frac{\beta}{\alpha}\right)^2\right\}dz\exp\left\{\frac{\beta^2}{\alpha^2}s\right\}ds. \tag{12}$$

Since we have

$$\int_{\mathbb{R}_0}e^{(iv+a)\frac{z}{\alpha}}\exp\left\{-\frac{1}{4s}\left(z-2s\frac{\beta}{\alpha}\right)^2\right\}\frac{dz}{\sqrt{2\pi 2s}}$$
$$=\exp\left\{i\frac{2s}{\alpha^2}(v\beta+va)-\frac{s}{\alpha^2}(v^2-a^2-2a\beta)\right\},$$

we obtain

$$\begin{aligned}(12)&=\frac{\delta\alpha}{2\sqrt{\pi}}\int_0^\infty\exp\left\{\left(\frac{\beta^2}{\alpha^2}-1\right)s\right\}s^{-\frac{3}{2}}\\&\quad\times\left(\exp\left\{i\frac{2s}{\alpha^2}(v\beta+va)-\frac{s}{\alpha^2}(v^2-a^2-2a\beta)\right\}-1\right)ds\\&=\frac{\delta\alpha}{2\sqrt{\pi}}\int_0^\infty s^{-\frac{3}{2}}\left(e^{ibs-M_1s}-e^{-M_2s}\right)ds\\&=\frac{\delta\alpha}{2\sqrt{\pi}}\int_0^\infty s^{-\frac{1}{2}}\left(ie^{-M_1s}\int_0^b e^{ius}du+\int_{M_1}^{M_2}e^{-us}du\right)ds\\&=\frac{\delta\alpha}{2\sqrt{\pi}}i\int_0^b\int_0^\infty e^{(iu-M_1)s}s^{-\frac{1}{2}}dsdu+\frac{\delta\alpha}{2\sqrt{\pi}}\int_{M_1}^{M_2}\int_0^\infty e^{-us}s^{-\frac{1}{2}}dsdu.\end{aligned}\tag{13}$$

On the other hand, we have

$$\int_{M_1}^{M_2}\int_0^\infty e^{-us}s^{-\frac{1}{2}}dsdu=\int_{M_1}^{M_2}\frac{\sqrt{\pi}}{\sqrt{u}}du=2\sqrt{\pi}(\sqrt{M_2}-\sqrt{M_1})$$

by $M_1, M_2 > 0$. As a result, using Lemma A.2, we obtain

$$\begin{aligned}(13)&=\frac{\delta\alpha}{2\sqrt{\pi}}\sqrt{2\pi}i\left\{\sqrt{\sqrt{M_1^2+b^2}-M_1}+i\left(\sqrt{\sqrt{M_1^2+b^2}+M_1}-\sqrt{2M_1}\right)\right\}\\&\quad+\frac{\delta\alpha}{2\sqrt{\pi}}2\sqrt{\pi}\left(\sqrt{M_2}-\sqrt{M_1}\right)\\&=\frac{\delta\alpha}{\sqrt{2}}\left\{i\sqrt{\sqrt{M_1^2+b^2}-M_1}-\sqrt{\sqrt{M_1^2+b^2}+M_1}+\sqrt{2M_2}\right\},\end{aligned}$$

from which (11) follows for any $v\in[0,\infty)$ and any $a\in(\frac{3}{2},2]$.

For $v \geq 0$, we see that (11) still holds for $a + 1$. To this end, it is enough to make sure that $M_1(v, a + 1)$ and $b(v, a + 1)$ remain nonnegative. In fact, we have

$$M_1(v, a + 1) = \frac{\alpha^2 - (a + 1 + \beta)^2}{\alpha^2} \geq 0$$

and

$$b(v, a + 1) = \frac{2(a + 1 + \beta)v}{\alpha^2} \geq 0$$

by Assumption 3.1. Similarly, (11) follows for the case of $(v, a) = (0, 1)$, since

$$M_1(0, 1) = \frac{\alpha^2 - (1 + \beta)^2}{\alpha^2} \geq \frac{6}{\alpha^2} > 0$$

and $b(0, 1) = 0$. □

Proof of Proposition 3.2

Noting that $0 \geq h > -1$ by Assumption 3.1 and Proposition 3.1, we have

$$\begin{aligned} \nu^{\mathbb{P}^*}(dx) &= (1 - \theta_x)\nu(dx) = (1 - h(e^x - 1))\nu(dx) = (1 + h)\nu(dx) - he^x\nu(dx) \\ &= \nu[\alpha, \beta, (1 + h)\delta](dx) + \nu[\alpha, 1 + \beta, -h\delta](dx) \end{aligned}$$

by (2). This completes the proof of Proposition 3.2. □

Proof of Proposition 3.3

To show Proposition 3.3, we start with the following lemma:

Lemma A.3 *We have*

$$\int_{\mathbb{R}_0} x\nu(dx) = \frac{\delta\beta}{\sqrt{\alpha^2 - \beta^2}}.$$

Proof The Sommerfeld integral representation (10) implies that

$$\int_{\mathbb{R}_0} x\nu(dx) = \frac{\delta}{4\pi}\int_{\mathbb{R}_0} z \exp\left\{\frac{\beta}{\alpha}z\right\}\int_0^\infty \exp\left\{-s - \frac{z^2}{4s}\right\}s^{-2}dsdz$$

$$
\begin{aligned}
&= \frac{\delta}{4\pi}\int_0^\infty \int_{\mathbb{R}_0} \frac{z}{\sqrt{2\pi 2s}} \exp\left\{-\frac{1}{4s}\left(z - 2s\frac{\beta}{\alpha}\right)^2\right\} dz\sqrt{2\pi 2s} \\
&\qquad \times \exp\left\{\frac{\beta^2}{\alpha^2}s\right\} s^{-2}e^{-s}ds \\
&= \frac{\delta\beta}{\sqrt{\pi}\alpha}\int_0^\infty \exp\left\{-\left(1-\frac{\beta^2}{\alpha^2}\right)s\right\} s^{-\frac{1}{2}}ds = \frac{\delta\beta}{\sqrt{\alpha^2-\beta^2}}.
\end{aligned}
$$

□

Note that we do not need Assumption 3.1 in the above proof. Now, we show Proposition 3.3. By Lemma A.3 and Proposition 3.2, we have

$$
\int_{\mathbb{R}_0}(iv+a)x\nu^{\mathbb{P}^*}(dx) = (iv+a)\left(\frac{(1+h)\delta\beta}{\sqrt{\alpha^2-\beta^2}} - \frac{h\delta(1+\beta)}{\sqrt{\alpha^2-(1+\beta)^2}}\right).
$$

Remark that $W(v, a; \alpha, 1+\beta, -h\delta)$ is well-defined and satisfies (11), since we have $M_1(v, a; \alpha, \beta+1) = M_1(v, a+1; \alpha, \beta) \geq 0$ and $b(v, a; \alpha, \beta+1) = b(v, a+1; \alpha, \beta) \geq 0$. (3) implies that

$$
\begin{aligned}
\phi_{T-t}(v-ia) &= \mathbb{E}_{\mathbb{P}^*}\left[e^{(iv+a)L_{T-t}}\right] \\
&= \mathbb{E}_{\mathbb{P}^*}\Bigg[\exp\Bigg\{(T-t)(iv+a)\mu^* \\
&\qquad + \int_{\mathbb{R}_0}(iv+a)x\widetilde{N}^{\mathbb{P}^*}([0,T-t],dx)\Bigg\}\Bigg] \\
&= \exp\Bigg\{(T-t)\Bigg((iv+a)\mu^* \\
&\qquad + \int_{\mathbb{R}_0}\left(e^{(iv+a)x}-1-(iv+a)x\right)\nu^{\mathbb{P}^*}(dx)\Bigg)\Bigg\} \\
&= \exp\left\{(T-t)(iv+a)\left(\mu^* - \frac{(1+h)\delta\beta}{\sqrt{\alpha^2-\beta^2}} + \frac{h\delta(1+\beta)}{\sqrt{\alpha^2-(1+\beta)^2}}\right)\right\} \\
&\quad \times \exp\Bigg\{(T-t)\Bigg(W(v,a;\alpha,\beta,(1+h)\delta) \\
&\qquad + W(v,a;\alpha,1+\beta,-h\delta)\Bigg)\Bigg\},
\end{aligned}
$$

from which Proposition 3.3 follows. □

Proof of Proposition 3.5

To see Proposition 3.5, we prepare one proposition and one lemma. In order to emphasize the parameters α, β, and δ, we write $M_1(v, a)$, M_2, and $b(v, a)$ as $M_1(v, a; \alpha, \beta)$, $M_2(\alpha, \beta)$, and $b(v, a; \alpha, \beta)$, respectively.

Proposition A.1 *For any* $v \in [0, \infty)$ *and any* $t \in [0, T)$, *we have*

$$|\phi_{T-t}(v - ia)| \le C(t)e^{-(T-t)\delta v},$$

where $C(t)$ *is given in* (8).

Proof Proposition 3.3 implies that

$$\begin{aligned}
&|\phi_{T-t}(v - ia)| \\
&= \left| \exp\left\{ (T-t)(iv + a)\left(\mu^* - \frac{(1+h)\delta\beta}{\sqrt{\alpha^2 - \beta^2}} + \frac{h\delta(1+\beta)}{\sqrt{\alpha^2 - (1+\beta)^2}} \right) \right\} \right. \\
&\quad \left. \times \exp\left\{ (T-t)\Big(W(v, a; \alpha, \beta, (1+h)\delta) + W(v, a; \alpha, 1+\beta, -h\delta) \Big) \right\} \right| \\
&= C(t)\exp\left\{ -(T-t)\frac{(1+h)\delta\alpha}{\sqrt{2}} \sqrt{\sqrt{M_1(v, a; \alpha, \beta)^2 + b(v, a; \alpha, \beta)^2} + M_1(v, a; \alpha, \beta)} \right\} \\
&\quad \times \exp\left\{ -(T-t)\frac{(-h)\delta\alpha}{\sqrt{2}} \sqrt{\sqrt{M_1(v, a; \alpha, 1+\beta)^2 + b(v, a; \alpha, 1+\beta)^2} + M_1(v, a; \alpha, 1+\beta)} \right\} \\
&\le C(t)\exp\left\{ -(T-t)(1+h)\delta\alpha\sqrt{M_1(v, a; \alpha, \beta)} \right\} \exp\left\{ -(T-t)(-h)\delta\alpha\sqrt{M_1(v, a; \alpha, 1+\beta)} \right\} \\
&= C(t)\exp\left\{ -(T-t)\delta\left((1+h)\sqrt{v^2 + \alpha^2 - (a+\beta)^2} + (-h)\sqrt{v^2 + \alpha^2 - (a+1+\beta)^2} \right) \right\} \\
&\le C(t)\exp\{-(T-t)\delta v\}.
\end{aligned}$$

Note that the last inequality follows from the fact that $\alpha^2 - (a+\beta)^2 > 0$ and $\alpha^2 - (a+1+\beta)^2 > 0$ hold by Assumption 3.1. □

Lemma A.4 *For any* $v \in [0, \infty)$ *and any* $a \in (\frac{3}{2}, 2]$,

$$|W(v, a+1) - W(v, a)| \le \sqrt{2}\delta\left(v + \sqrt{\alpha^2 - (a+\beta)^2 + 2(a+1+\beta)^2} \right)$$

holds.

Proof Denoting $M_1' := M_1(v, a+1)$, $b' := b(v, a+1)$, $M_1 := M_1(v, a)$, and $b := b(v, a)$, for short, we have

$$|W(v, a+1) - W(v, a)|$$

$$
\begin{aligned}
&= \frac{\delta\alpha}{\sqrt{2}}\left|i\left(\sqrt{\sqrt{M_1'^2+b'^2}-M_1'}-\sqrt{\sqrt{M_1^2+b^2}-M_1}\right)\right.\\
&\quad\left.-\sqrt{\sqrt{M_1'^2+b'^2}+M_1'}+\sqrt{\sqrt{M_1^2+b^2}+M_1}\right|\\
&\le \delta\alpha\sqrt{\sqrt{M_1'^2+b'^2}+\sqrt{M_1^2+b^2}}. \qquad (14)
\end{aligned}
$$

Since $a+\beta>0$, we have

$$
M_1 - M_1' = \frac{1}{\alpha^2}\left((a+1+\beta)^2-(a+\beta)^2\right) > 0
$$

and

$$
b'^2 - b^2 = \frac{4v^2}{\alpha^4}\left((a+1+\beta)^2-(a+\beta)^2\right) > 0,
$$

which imply that

$$
\begin{aligned}
(14) &\le \delta\alpha\sqrt{2\sqrt{M_1^2+b'^2}} = \sqrt{2}\delta\sqrt[4]{(v^2+\alpha^2-(a+\beta)^2)^2+4v^2(a+1+\beta)^2}\\
&= \sqrt{2}\delta\sqrt[4]{v^4+2v^2(\alpha^2-(a+\beta)^2+2(a+\beta+1)^2)+(\alpha^2-(a+\beta)^2)^2}. \qquad (15)
\end{aligned}
$$

Setting

$$
\begin{cases}
p := \alpha^2-(a+\beta)^2+2(a+\beta+1)^2,\\
q := p^2-(\alpha^2-(a+\beta)^2)^2,
\end{cases}
$$

we have $p>0$ and $q>0$ for any $a\in(\frac{3}{2},2]$ by Assumption 3.1 and

$$
(15) = \sqrt{2}\delta\sqrt[4]{(v^2+p)^2-q} \le \sqrt{2}\delta\sqrt{v^2+p} \le \sqrt{2}\delta(v+\sqrt{p}).
$$

This completes the proof of Lemma A.4. □

Proof of Proposition 3.5. Firstly, Lemma A.4 implies that

$$
\begin{aligned}
&\left|\frac{1}{\pi}\int_w^\infty K^{-iv-a+1}\Big(W(v,a+1)-W(v,a)-W(0,1)\Big)\frac{\phi_{T-t}(iv-a)S_{t-}^{iv+a}}{(iv+a)(iv+a-1)}dv\right|\\
&\le \frac{1}{\pi}\int_w^\infty \left|K^{-iv-a+1}\right|\Big(|W(v,a+1)-W(v,a)|+|W(0,1)|\Big)\\
&\qquad\times\left|\frac{\phi_{T-t}(iv-a)S_{t-}^{iv+a}}{(iv+a)(iv+a-1)}\right|dv
\end{aligned}
$$

$$\leq \frac{\delta K^{-a+1}}{\pi}\int_w^\infty \left(\sqrt{2}(v+\sqrt{p})+\sqrt{2}\right)\left|\frac{\phi_{T-t}(iv-a)S_{t-}^{iv+a}}{(iv+a)(iv+a-1)}\right|dv, \tag{16}$$

where p is defined in the proof of Lemma A.4. Remark that the last inequality in (16) holds since

$$|W(0,1)| = \delta\alpha\left(\sqrt{M_1(0,1)}-\sqrt{M_2}\right) = \delta\alpha\left(\sqrt{M_2-\frac{1+2\beta}{\alpha^2}}-\sqrt{M_2}\right)$$
$$\leq \delta\alpha\sqrt{-\frac{1+2\beta}{\alpha^2}} \leq \sqrt{2}\delta$$

by Assumption 3.1. Now, note that

$$|(iv+a-1)(iv+a)| = \sqrt{(a^2-a-v^2)^2+(2a-1)^2v^2}$$
$$= \sqrt{v^4+(2a^2-2a+1)v^2+(a^2-a)^2} \geq v^2.$$

Thus, Proposition A.1 implies that

$$\left|\frac{\phi_{T-t}(iv-a)S_{t-}^{iv+a}}{(iv+a)(iv+a-1)}dv\right| \leq \frac{S_{t-}^a C(t)}{v^2}e^{-(T-t)\delta v}.$$

As a result, noting that $w > 1$, we obtain

$$(16) \leq \frac{\delta K^{-a+1}S_{t-}^a C(t)}{\pi}\int_w^\infty \left(\sqrt{2}(v+\sqrt{p})+\sqrt{2}\right)\frac{1}{v^2}e^{-(T-t)\delta v}dv$$
$$\leq \frac{\delta K^{-a+1}S_{t-}^a C(t)}{\pi}\int_w^\infty \left(2\sqrt{2}+\sqrt{2p}\right)e^{-(T-t)\delta v}dv$$
$$= \frac{K^{-a+1}S_{t-}^a C(t)}{\pi}\frac{\sqrt{2}(2+\sqrt{p})}{T-t}e^{-(T-t)\delta w}.$$

This completes the proof of Proposition 3.5. □

Acknowledgements This work was supported by JSPS KAKENHI Grant Numbers 15K04936 and 17K13764.

References

1. Arai T, Imai Y (2017) A closed-form representation of mean-variance hedging for additive processes via Malliavin calculus, preprint. Available at https://arxiv.org/abs/1702.07556
2. Arai T, Imai Y, Suzuki R (2016) Numerical analysis on local risk-minimization for exponential Lévy models. Int J Theor Appl Financ 19:1650008
3. Arai T, Suzuki R (2015) Local risk-minimization for Lévy markets. Int J Financ Eng 2:1550015
4. Barndorff-Nielsen OE (1995) Normal inverse Gaussian processes and the modelling of stock returns. Aarhus Universitet, Department of Theoretical Statistics, Aarhus
5. Barndorff-Nielsen OE (1997) Processes of normal inverse Gaussian type. Finance Stochast 2:41–68
6. Barndorff-Nielsen OE (1997) Normal inverse Gaussian distributions and stochastic volatility modelling. Scand J Stat 24:1–13
7. Benth FE, Šaltytė-Benth J (2004) The normal inverse Gaussian distribution and spot price modelling in energy markets. Int J Theor Appl Financ 7:177–192
8. Carr P, Madan D (1999) Option valuation using the fast Fourier transform. J Comput Financ 2:61–73
9. Cont R, Tankov P (2004) Financial modelling with jump process. Chapman & Hall, London
10. Delong Ł, Imkeller P (2010) On Malliavin's differentiability of BSDEs with time delayed generators driven by Brownian motions and Poisson random measures. Stoch Process Appl 120:1748–1775
11. Rydberg TH (1997) The normal inverse Gaussian Lévy process: simulation and approximation. Commun Stat Stoch Models 13:887–910
12. Rydberg TH (1997) A note on the existence of unique equivalent martingale measures in a Markovian setting. Financ Stoch 1:251–257
13. Schoutens W (2003) Lévy process in finance: pricing financial derivatives. Wiley, Hoboken
14. Schweizer M (2001) A guided tour through quadratic hedging approaches. In: Jouini E, Cvitanic J, Musiela M (eds) Option pricing, interest rates and risk management. Cambridge University Press, Cambridge, pp 538–574
15. Schweizer M (2008) Local risk-minimization for multidimensional assets and payment streams. Banach Cent Publ 83:213–229
16. Solé JL, Utzet F, Vives J (2007) Canonical Lévy process and Malliavin calculus. Stoch Process Appl 117:165–187

Second-Order Evolution Problems with Time-Dependent Maximal Monotone Operator and Applications

C. Castaing, M. D. P. Monteiro Marques, and P. Raynaud de Fitte

Abstract We consider at first the existence and uniqueness of solution for a general second-order evolution inclusion in a separable Hilbert space of the form

$$0 \in \ddot{u}(t) + A(t)\dot{u}(t) + f(t, u(t)),\ t \in [0, T]$$

where $A(t)$ is a time dependent with Lipschitz variation maximal monotone operator and the perturbation $f(t, .)$ is boundedly Lipschitz. Several new results are presented in the sense that these second-order evolution inclusions deal with time-dependent maximal monotone operators by contrast with the classical case dealing with some special fixed operators. In particular, the existence and uniqueness of solution to

$$0 = \ddot{u}(t) + A(t)\dot{u}(t) + \nabla\varphi(u(t)),\ t \in [0, T]$$

where $A(t)$ is a time dependent with Lipschitz variation single-valued maximal monotone operator and $\nabla\varphi$ is the gradient of a smooth Lipschitz function φ are stated. Some more general inclusion of the form

$$0 \in \ddot{u}(t) + A(t)\dot{u}(t) + \partial\Phi(u(t)),\ t \in [0, T]$$

Work partially supported by Fundação para a Ciência e a Tecnologia, UID/MAT/04561/2013.

JEL Classification: C61, C73

Mathematics Subject Classification (2010): 34A60, 34B15

C. Castaing (✉)
IMAG, Université de Montpellier, Montpellier, France

M. D. P. Monteiro Marques
CMAF-CIO, Departamento de Matemática, Faculdade de Ciências da Universidade de Lisboa, Lisbon, Portugal
e-mail: mdmarques@fc.ul.pt

P. Raynaud de Fitte
Laboratoire de Mathématiques Raphaël Salem, Normandie Université, Rouen, France
e-mail: prf@univ-rouen.fr

S. Kusuoka, T. Maruyama (eds.), *Advances in Mathematical Economics*, Advances in Mathematical Economics 22, https://doi.org/10.1007/978-981-13-0605-1_2

where $\partial\Phi(u(t))$ denotes the subdifferential of a proper lower semicontinuous convex function Φ at the point $u(t)$ is provided via a variational approach. Further results in second-order problems involving both absolutely continuous in variation maximal monotone operator and bounded in variation maximal monotone operator, $A(t)$, with perturbation $f : [0, T] \times H \times H$ are stated. Second- order evolution inclusion with perturbation f and Young measure control ν_t

$$\begin{cases} 0 \in \ddot{u}_{x,y,\nu}(t) + A(t)\dot{u}_{x,y,\nu}(t) + f(t, u_{x,y,\nu}(t)) + \mathrm{bar}(\nu_t), \ t \in [0, T] \\ u_{x,y,\nu}(0) = x, \ \dot{u}_{x,y,\nu}(0) = y \in D(A(0)) \end{cases}$$

where $\mathrm{bar}(\nu_t)$ denotes the barycenter of the Young measure ν_t is considered, and applications to optimal control are presented. Some variational limit theorems related to convex sweeping process are provided.

Keywords Bolza control problem · Lipschitz mapping · Maximal monotone operators · Pseudo-distance · Subdifferential · Viscosity · Young measures

Article type: Research Article
Received: March 15, 2018
Revised: March 30, 2018

1 Introduction

Let H be a separable Hilbert space. In this paper, we are mainly interested in the study of the perturbed evolution problem

$$0 \in \ddot{u}(t) + A(t)\dot{u}(t) + \partial\Phi(u(t)), \ t \in [0, T]$$

where $\partial\Phi(u(t))$ denotes the subdifferential of a proper lower semicontinuous convex function Φ at the point $u(t)$, $A(t) : D(A(t)) \to 2^H$ is a maximal monotone operator in the Hilbert space H for every $t \in [0, T]$, and the dependence $t \mapsto A(t)$ has *Lipschitz variation*, in the sense that there exists $\alpha \geq 0$ such that

$$\mathrm{dis}(A(t), A(s)) \leq \alpha(t - s), \ \forall s, t \in [0, T] \ (s \leq t)$$

dis(., .) being the *pseudo-distance* between maximal monotone operators (m.m.o.) defined by A. A. Vladimirov [53] as

$$\mathrm{dis}(A, B) = \sup \left\{ \frac{\langle y - \hat{y}, \hat{x} - x \rangle}{1 + ||y|| + ||\hat{y}||} : x \in D(A), y \in Ax, \hat{x} \in D(B), \hat{y} \in B\hat{x} \right\}$$

for m.m.o. A and B with domains $D(A)$ and $D(B)$, respectively; the dependence $t \mapsto A(t)$ has *absolutely continuous variation*, in the sense that there exists $\beta \in W^{1,1}([0, T])$ such that

$$\text{dis}(A(t), A(s)) \leq |\beta(t) - \beta(s)|, \ \forall t, s \in [0, T],$$

the dependence $t \mapsto A(t)$ has *bounded variation* in the sense that there exists a function $r : [0, T] \to [0, +\infty[$ which is continuous on $[0, T[$ and nondecreasing with $r(T) < +\infty$ such that

$$\text{dis}(A(t), A(s)) \leq dr(]s, t]) = r(t) - r(s) \ \text{ for } \ 0 \leq s \leq t \leq T$$

The paper is organized as follows. Section 2 contains some definitions, notation and preliminary results. In Sect. 3, we recall and summarize (Theorem 3.2) the existence and uniqueness of solution for a general second-order evolution inclusion in a separable Hilbert space of the form

$$0 \in \ddot{u}(t) + A(t)\dot{u}(t) + f(t, u(t)), \ t \in [0, T]$$

where $A(t)$ is a time dependent with Lipschitz variation maximal monotone operator and the perturbation $f(t, .)$ is *dt-boundedly Lipschitz* (short for *dt-integrably Lipschitz on bounded sets*). At this point, Theorem 3.2 and its corollaries are new results in the sense that these second-order evolution inclusions deal with time-dependent maximal monotone operators by contrast with the classical case dealing with some special fixed operators; cf. Attouch et al. [4], Paoli [43], and Schatzman [48]. In particular, the existence and uniqueness of solution, based on Corollary 3.2, to

$$0 = \ddot{u}(t) + A(t)\dot{u}(t) + \nabla\varphi(u(t)), \ t \in [0, T]$$

where $A(t)$ is a time dependent with Lipschitz variation single-valued maximal monotone operator and $\nabla\varphi$ is the gradient of a smooth Lipschitz function φ, have some importance in mechanics [40], which may require a more general evolution inclusion of the form

$$0 \in \ddot{u}(t) + A(t)\dot{u}(t) + \partial\Phi(u(t)), \ t \in [0, T]$$

where $\partial\Phi(u(t))$ denotes the subdifferential of a proper lower semicontinuous convex function Φ at the point $u(t)$.

We provide (Proposition 3.1) the existence of a generalized $W_{BV}^{1,1}([0, T], H)$ solution to the second-order inclusion $0 \in \ddot{u}(t) + A(t)\dot{u}(t) + \partial\Phi(u(t))$ which enjoys several regularity properties. The result is similar to that of Attouch et al. [4], Paoli [43], and Schatzman [48] with different hypotheses and a different method that is essentially based on Corollary 3.2 and the tools given in [22, 23, 27] involving the Young measures and biting convergence [9, 22, 32]. By $W^{1,1}_{BV}([0,T],H)$, we denote the space of all absolutely continuous mappings $y : [0,T] \to H$ such that $\dot{y}$ are BV. Further results on second-order problems involving both the absolutely continuous in variation maximal monotone operators and the bounded in variation maximal monotone operator $A(t)$ with perturbation $f : [0,T] \times H \times H$ are stated.

Finally, in Sect. 4, we present several applications in optimal control in a new setting such as Bolza relaxation problem, dynamic programming principle, viscosity in evolution inclusion driven by a Lipschitz variation maximal monotone operator $A(t)$ with Lipschitz perturbation f, and Young measure control ν_t

$$\begin{cases} 0 \in \ddot{u}_{x,y,\nu}(t) + A(t)\dot{u}_{x,y,\nu}(t) + f(t, u_{x,y,\nu}(t)) + \text{bar}(\nu_t), \ t \in [0,T] \\ u_{x,y,\nu}(0) = x, \ \dot{u}_{x,y,\nu}(0) = y \in D(A(0)) \end{cases}$$

where $\text{bar}(\nu_t)$ denotes the barycenter of the Young measure ν_t in the same vein as in Castaing-Marques-Raynaud de Fitte [25] dealing with the sweeping process. At this point, the above second-order evolution inclusion contains the evolution problem associated with the sweeping process by a closed convex Lipschitzian mapping $C : [0,T] \to \text{cc}(H)$

$$\begin{cases} 0 \in \ddot{u}(t) + N_{C(t)}(\dot{u}(t)) + f(t, u(t)) + \text{bar}(\nu_t), \ t \in [0,T] \\ u(0) = u_0, \ \dot{u}(0) = \dot{u}_0 \in C(0) \end{cases}$$

(where $\text{cc}(H)$ denotes the set of closed convex subsets of H) by taking $A(t) = \partial\Psi_{C(t)}$ and noting that if $C(t)$ is a closed convex moving set in H, then the subdifferential of its indicator function is $A(t) = \partial\Psi_{C(t)} = N_{C(t)}$, the outward normal cone operator. Since for all $s, t \in [0,T]$

$$\text{dis}\big(A(t), A(s)\big) = \mathcal{H}\big(C(t), C(s)\big),$$

where $\mathcal{H}$ denotes the Hausdorff distance; it follows that our study of these time-dependent maximal monotone operators includes as special cases some related results for evolution problems governed by sweeping process of the form

$$0 \in \ddot{u}(t) + N_{C(t)}(\dot{u}(t)) + f(t, u(t)), \ t \in [0,T].$$

Since now sweeping process has found applications in several fields in particular to economics [29, 31, 35], we present also some variational limit theorems related to convex sweeping process; see [1, 3, 34] and the references therein.

There is a vast literature on evolution inclusions driven by the sweeping process and the subdifferential operators. See [2, 5, 6, 10, 17, 18, 20, 21, 25, 26, 28, 30, 37, 39–41, 45, 47, 49–52] and the references therein. We refer to [9, 12, 13, 54] for the study of maximal monotone operators.

2 Notation and Preliminaries

In the whole paper, $I := [0, T]$ $(T > 0)$ is an interval of $\mathbb{R}$, and H is a real Hilbert space whose scalar product will be denoted by $\langle \cdot, \cdot \rangle$ and the associated norm by $\| \cdot \|$. $\mathscr{L}([0, T])$ is the Lebesgue σ-algebra on $[0, T]$, and $\mathscr{B}(H)$ is the σ-algebra of Borel subsets of H. We will denote by $\overline{\mathbf{B}}_H(x_0, r)$ the closed ball of H of center x_0 and radius $r > 0$ and by $\overline{\mathbf{B}}_H$ its closed unit ball. $\mathrm{C}(I, H)$ denotes the Banach space of all continuous mappings $u : I \to H$ equipped with the norm $\|u\|_C = \max\limits_{t \in I} \|u(t)\|$. For $q \in [1, +\infty[$, $L^q_H([0, T], dt)$ is the space of (classes of) measurable $u : [0, T] \to H$, with the norm $\|u(\cdot)\|_q = (\int_0^T \|u(t)\|^q dt)^{\frac{1}{q}}$, and $L^\infty_H([0, T], dt)$ is the space of (classes of) measurable essentially bounded $u : [0, T] \to H$ equipped with $\|.\|_\infty$.

If E is a Banach space and E^* its topological dual, we denote by $\sigma(E, E^*)$ the weak topology on E and by $\sigma(E^*, E)$ the weak star topology on E^*. For any $C \subset E$, we denote by $\delta^*(., C)$ the support function of C, i.e.

$$\delta^*(x^*, C) = \sup_{x \in C} \langle x^*, x \rangle \, , \forall x^* \in E^*.$$

A set-valued map $A : D(A) \subset H \to 2^H$ is monotone if $\langle y_1 - y_2, x_1 - x_2 \rangle \geq 0$ whenever $x_i \in D(A)$ and $y_i \in A(x_i)$, $i = 1, 2$. A monotone operator A is maximal if A is not contained properly in any other monotone operator, that is, for all $\lambda > 0$, $R(I_H + \lambda A) = H$, with $R(A) = \bigcup \{Ax, \ x \in D(A)\}$ the range of A and I_H the identity mapping of H. In the whole paper, $I := [0, T]$ $(T > 0)$ is an interval of $\mathbb{R}$, and H is a real Hilbert space whose scalar product will be denoted by $\langle \cdot, \cdot \rangle$ and the associated norm by $\| \cdot \|$. Let $A : D(A) \subset H \to 2^H$ be a set-valued map. We say that A is monotone, if $\langle y_1 - y_2, x_1 - x_2 \rangle \geq 0$ whenever $x_i \in \mathscr{D}(A)$ and $y_i \in A(x_i)$, $i = 1, 2$. If $\langle y_1 - y_2, x_1 - x_2 \rangle = 0$ implies that $x_1 = x_2$, we say that A is strictly monotone. A monotone operator A is said to be maximal if A could not be contained properly in any other monotone operator.

If A is a maximal monotone operator, then, for every $x \in D(A)$, $A(x)$ is nonempty closed and convex. So the set $A(x)$ contains an element of minimum norm (the projection of the origin on the set $A(x)$). This unique element is denoted by $A^0(x)$. Therefore $A^0(x) \in A(x)$ and $\|A^0(x)\| = \inf_{y \in A(x)} \|y\|$. Moreover the set $\overline{D(A)}$ is convex.

For $\lambda > 0$, we define the following well-known operators:

$$J^A_\lambda = (I + \lambda A)^{-1} \text{ (the resolvent of } A),$$

$$A_\lambda = \frac{1}{\lambda}(I - J^A_\lambda) \text{(the Yosida approximation of } A).$$

The operators J^A_λ and A_λ are defined on all of H. For the terminology of maximal monotone operators and more details, we refer the reader to [9, 13], and [54].

Let $A : D(A) \subset H \to 2^H$ and $B : D(B) \subset H \to 2^H$ be two maximal monotone operators, and then we denote by $\mathrm{dis}(A, B)$ the pseudo-distance between

A and B defined by A. A. Vladimirov [53] as

$$\operatorname{dis}(A,B)=\sup\left\{\frac{\langle y-y',x'-x\rangle}{1+\|y\|+\|y'\|}:\ x\in D(A),\ y\in Ax,\ x'\in D(B),\ y'\in Bx'\right\}.$$

Our main results are established under the following hypotheses on the operator A:

$(H1)$ The mapping $t\mapsto A(t)$ has Lipschitz variation, in the sense that there exists $\alpha\geq 0$ such that

$$\operatorname{dis}(A(t),A(s))\leq \alpha(t-s),\ \forall s,t\in[0,T]\ (s\leq t).$$

$(H2)$ There exists a nonnegative real number c such that

$$\|A^0(t,x)\|\leq c(1+\|x\|)\ \text{ for }\ t\in[0,T],\ x\in D(A(t)).$$

We recall some elementary lemmas, and we refer to [38] for the proofs.

Lemma 2.1 *Let A and B be maximal monotone operators. Then*

(1) $\operatorname{dis}(A,B)\in[0,+\infty]$, $\operatorname{dis}(A,B)=\operatorname{dis}(B,A)$ *and* $\operatorname{dis}(A,B)=0$ *iff* $A=B$.
(2) $\|x-Proj(x,\overline{D(B)})\|\leq \operatorname{dis}(A,B)$ *for* $x\in\overline{D(A)}$.
(3) $\mathcal{H}(D(A),D(B))\leq \operatorname{dis}(A,B)$.

Lemma 2.2 *Let A be a maximal monotone operator. If $x,y\in H$ are such that*

$$\langle A^0(z)-y,z-x\rangle\geq 0\ \ \forall z\in D(A),$$

then $x\in D(A)$ and $y\in A(x)$.

Lemma 2.3 *Let A_n $(n\in\mathbb{N})$ and A be maximal monotone operators such that* $\operatorname{dis}(A_n,A)\to 0$. *Suppose also that $x_n\in D(A_n)$ with $x_n\to x$ and $y_n\in A_n(x_n)$ with $y_n\to y$ weakly for some $x,y\in H$. Then $x\in D(A)$ and $y\in A(x)$.*

Lemma 2.4 *Let A and B be maximal monotone operators. Then*

(1) *for $\lambda>0$ and $x\in D(A)$*

$$\|x-J_\lambda^B(x)\|\leq\lambda\|A^0(x)\|+\operatorname{dis}(A,B)+\sqrt{\lambda\big(1+\|A^0(x)\|\big)\operatorname{dis}(A,B)}.$$

(2) *For $\lambda>0$ and $x,x'\in H$*

$$\|J_\lambda^A(x)-J_\lambda^B(x')\|^2\leq\|x-x'\|^2+2\lambda\big(1+\|A_\lambda(x)\|+\|B_\lambda(x')\|\big)\operatorname{dis}(A,B).$$

(3) *For $\lambda>0$ and $x,x'\in H$*

$$\|A_\lambda(x)-B_\lambda(x')\|^2\leq\frac{1}{\lambda^2}\|x-x'\|^2+\frac{2}{\lambda}\big(1+\|A_\lambda(x)\|+\|B_\lambda(x')\|\big)\operatorname{dis}(A,B).$$

3 Second-Order Evolution Problems Involving Time-Dependent Maximal Monotone Operators

In the sequel, H is a separable Hilbert space. For the sake of completeness, we summarize and state the following result. We say that a function $f = f(t, x)$ is *dt-boundedly Lipschitz* (short for *dt-integrably Lipschitz on bounded sets*) if, for every $R > 0$, there is a nonnegative dt-integrable function $\lambda_R \in L^1([0, T], \mathbb{R}; dt)$ such that, for all $t \in [0, T]$

$$\|f(t,x) - f(t,y)\| \leq \lambda_R(t)||x - y||, \; \forall x, y \in \overline{\mathbf{B}}(0, R).$$

Theorem 3.1 *Let for every $t \in [0, T]$, $A(t) : D(A(t)) \subset H \to 2^H$ be a maximal monotone operator satisfying*

$(H1)$ *there exists a real constant $\alpha \geq 0$ such that*

$$\text{dis}(A(t), A(s)) \leq \alpha(t - s) \;\; \textit{for} \;\; 0 \leq s \leq t \leq T.$$

$(H2)$ *there exists a nonnegative real number c such that*

$$\|A^0(t,x)\| \leq c(1 + \|x\|), t \in [0, T], x \in D(A(t))$$

Let $f : [0, T] \times H \to H$ satisfying the linear growth condition

$(H3)$ *there exists a nonnegative real number M such that*

$$\|f(t,x)\| \leq M(1 + \|x\|) \;\; \textit{for} \;\; t \in [0, T], \; x \in H.$$

and assume that $f(., x)$ is dt-integrable for every $x \in H$. Assume also that f is dt-boundedly Lipschitz, as above.

Then for all $u_0 \in D(A(0))$, the problem

$$-\frac{du}{dt}(t) \in A(t)u(t) + f(t, u(t)) \; dt - \text{a.e.}\, t \in [0, T], \;\; u(0) = u_0$$

has a unique Lipschitz solution with the property: $||u(t) - u(\tau)|| \leq K \max\{1, \alpha\}|t - \tau|$ for all $t, \tau \in [0, T]$ for some constant $K \in]0, \infty[$.

Proof See [7, Theorem 3.1 and Theorem 3.3].

Theorem 3.2 *Let for every $t \in [0, T]$, $A(t) : D(A(t)) \subset H \to 2^H$ be a maximal monotone operator satisfying*

$(H1)$ *there exists a real constant $\alpha \geq 0$ such that*

$$\text{dis}(A(t), A(s)) \leq \alpha(t - s) \;\; \textit{for} \;\; 0 \leq s \leq t \leq T.$$

(*H*2) *there exists a nonnegative real number c such that*

$$\|A^0(t,x)\| \le c(1+\|x\|), t \in [0,T], x \in D(A(t))$$

Let $f : [0,T] \times H \to H$ *satisfying the linear growth condition:*

(*H*3) *there exists a nonnegative real number M such that*

$$\|f(t,x)\| \le M(1+\|x\|) \text{ for } t \in [0,T],\ x \in H.$$

and assume that $f(.,x)$ *is* dt*-integrable for every* $x \in H$*. Assume also that* f *is* dt*-boundedly Lipschitz.*

Then the second-order evolution inclusion

$$(\mathcal{S}_1)\begin{cases} 0 \in \ddot{u}(t) + A(t)\dot{u}(t) + f(t,u(t)),\ t \in [0,T] \\ u(0) = u_0,\ \dot{u}(0) = \dot{u}_0 \in D(A(0)) \end{cases}$$

admits a unique solution $u \in W_H^{2,\infty}([0,T], dt)$.

Proof The proof is a careful application of Theorem 3.1. In the new variables $X = (x, \dot{x})$, let us set for all $t \in I$

$$B(t)X = \{0\} \times A(t)\dot{x},\ \ g(t,X) = (-\dot{x}, f(t,x)).$$

For any $u \in W^{2,\infty}(I, H; dt)$, define $X(t) = (u(t), \frac{du}{dt}(t))$ and $\dot{X}(t) = \frac{dX}{dt}(t)$. Then the evolution inclusion $(\mathcal{S}_1)$ can be written as a first-order evolution inclusion associated with the Lipschitz maximal monotone operator $B(t)$ and the locally Lipschitz perturbation g:

$$\begin{cases} 0 \in \dot{X}(t) + B(t)X(t) + g(t,X(t)),\ t \in [0,T] \\ X(0) = (u_0, \dot{u}_0) \in H \times D(A(0)). \end{cases}$$

So the existence and uniqueness solution to the second-order evolution inclusion under consideration follows from Theorem 3.1.

There are some useful corollaries to Theorem 3.2.

Corollary 3.1 *Assume that for every* $t \in [0,T]$, $A(t) : H \to H$ *is a single-valued maximal monotone operator satisfying* (*H*1) *and* (*H*2). *Let* $f : [0,T] \times H \to H$ *be as in Theorem* 3.2. *Then the second-order evolution equation*

$$\begin{cases} 0 = \ddot{u}(t) + A(t)\dot{u}(t) + f(t,u(t)),\ t \in [0,T] \\ u(0) = u_0,\ \dot{u}(0) = \dot{u}_0 \end{cases}$$

admits a unique solution $u \in W_H^{2,\infty}([0,T])$.

Corollary 3.2 *Assume that for every $t \in [0, T]$, $A(t) : H \to H$ is a single-valued maximal monotone operator satisfying $(H1)$ and $(H2)$. Assume further that $A(t)$ satisfies*

(i) *$(t, x) \mapsto A(t)x$ is a Caratheodory mapping, that is, $t \mapsto A(t)x$ is Lebesgue measurable on $[0, T]$ for each fixed $x \in H$, and $x \mapsto A(t)x$ is continuous on H for each fixed $t \in [0, T]$,*
(ii) *$\langle A(t)x, x\rangle \geq \gamma||x||^2$, for all $(t, x) \in [0, T] \times H$, for some $\gamma > 0$.*

Let $\varphi \in C^1(H, \mathbb{R})$ be Lipschitz and such that $\nabla\varphi$ is locally Lipschitz. Then the evolution equation

$$(\mathscr{S}_2) \begin{cases} 0 = \ddot{u}(t) + A(t)\dot{u}(t) + \nabla\varphi(u(t)), \ t \in [0, T] \\ u(0) = u_0, \dot{u}(0) = \dot{u}_0 \end{cases}$$

admits a unique solution $u \in W^{2,\infty}([0, T], H; dt)$; moreover, u satisfies the energy estimate

$$\varphi(u(t)) - \frac{1}{2}||\dot{u}(t)||^2 \leq \varphi(u(0)) - \frac{1}{2}||\dot{u}(t)||^2 - \gamma \int_0^t ||\dot{u}(s)||^2 ds, \ t \in [0, T].$$

Proof Existence and uniqueness of solution follows from Theorem 3.2 or Corollary 3.1. The energy estimate is quite standard. Multiplying the equation by $\dot{u}(t)$ and applying the usual chain rule formula gives for all $t \in [0, T]$

$$\frac{d}{dt}\left(\varphi(u(t)) + \frac{1}{2}||\dot{u}(t)||^2\right) = -\langle A(t)\dot{u}(t), \dot{u}(t)\rangle.$$

By (i) and (ii) and by integrating on $[0, t]$, we get the required inequality

$$\begin{aligned}\varphi(u(t)) + \frac{1}{2}||\dot{u}(t)||^2 &= \varphi(u(0)) + \frac{1}{2}||\dot{u}(0)||^2 - \int_0^t \langle A(s)\dot{u}(s), \dot{u}(s)\rangle ds \\ &\leq \varphi(u(0)) + \frac{1}{2}||\dot{u}(0)||^2 - \gamma \int_0^t ||\dot{u}(s)||^2 ds, \ t \in [0, T],\end{aligned}$$

which completes the proof.

It is worth mentioning that the uniqueness of the solution to the equation $(\mathscr{S}_1)$ is quite important in applications, such as models in mechanics, since it contains the classical inclusion of the form

$$0 \in \ddot{u}(t) + \partial\Phi(\dot{u}(t)) + \nabla g(u(t))$$

where $\partial\Phi$ is the subdifferential of the proper lower semicontinuous convex function Φ and g is of class C^1 and ∇g is Lipschitz continuous on bounded sets. We also note that the uniqueness of the solution to the equation $(\mathscr{S}_2)$ and its energy estimate allow to recover a classical result in the literature dealing with finite dimensional space H and $A(t) = \gamma I_H$, $t \in [0, T]$, where I_H is the identity mapping in H. See Attouch et al. [4]. The energy estimate for the solution of

$$\begin{cases} 0 = \ddot{u}(t) + \gamma \dot{u}(t) + \nabla\varphi(u(t)), \ t \in I \\ u(0) = u_0, \dot{u}(0) = \dot{u}_0 \end{cases}$$

is then

$$\varphi(u(t)) + \frac{1}{2}||\dot{u}(t)||^2 = \varphi(u_0) + \frac{1}{2}||\dot{u}_0||^2 - \gamma \int_0^t ||\dot{u}(s)||^2 ds.$$

Actually the dynamical system $(\mathscr{S}_1)$ given in Theorem 3.2 has been intensively studied by many authors in particular cases. See Attouch et al. [4] dealing with the inclusion

$$0 \in \ddot{u}(t) + \gamma \dot{u}(t) + \partial\varphi(u(t))$$

and Paoli [43] and Schatzman [48] dealing with the second-order dynamical systems of the form

$$0 \in \ddot{u}(t) + \partial\varphi(u(t))$$

and

$$0 \in \ddot{u}(t) + A\dot{u}(t) + \partial\varphi(u(t))$$

where A is a positive autoadjoint operator. The existence and uniqueness of solutions in $(\mathscr{S}_2)$ are of some importance since they allow to obtain the existence of at least a $W_{BV}^{1,1}([0, T], H)$ solution with conservation of energy (see Proposition 3.1 below) for a second-order evolution inclusion of the form

$$(\mathscr{S}_3) \begin{cases} 0 \in \ddot{u}(t) + A(t)\dot{u}(t) + \partial\Phi(u(t), \ t \in I \\ u(0) = u_0 \in \text{dom}\,\Phi, \dot{u}(0) = \dot{u}_0 \in D(A(0)) \end{cases}$$

where $\partial\Phi$ is the subdifferential of a proper convex lower semicontinuous function; the energy estimate is given by

$$\Phi(u(t)) + \frac{1}{2}||\dot{u}(t)||^2 = \Phi(u(0)) + \frac{1}{2}||\dot{u}(0)||^2 - \int_0^t \langle A(s)\dot{u}(s), \dot{u}(s)\rangle ds.$$

Taking into account these considerations, we will provide the existence of a generalized solution to the second-order inclusion of the form

$$0 \in \ddot{u}(t) + A(t)\dot{u}(t) + \partial\phi(u(t))$$

which enjoy several regular properties. The result is similar to that of Attouch et al. [4], Paoli [43], and Schatzman [48] with different hypotheses and a different method that is essentially based on Corollary 3.2 and the tools given in [22, 23, 27] involving the Young measures [9, 32] and biting convergence.

Let us recall a useful Gronwall-type lemma [21].

Lemma 3.5 (A Gronwall-like inequality.) *Let $p, q, r : [0, T] \to [0, \infty[$ be three nonnegative Lebesgue integrable functions such that for almost all $t \in [0, T]$*

$$r(t) \leq p(t) + q(t) \int_0^t r(s)\, ds.$$

Then

$$r(t) \leq p(t) + q(t) \int_0^t \left[p(s) \exp\left(\int_s^t q(\tau)\, d\tau \right) \right] ds$$

for all $t \in [0, T]$.

Proposition 3.1 *Assume that $H = \mathbb{R}^d$ and that, for every $t \in [0, T]$, $A(t) : H \to H$ is single-valued maximal monotone satisfying*

$(H1)$ *there exists $\alpha > 0$ such that*

$$\text{dis}(A(t), A(s)) \leq \alpha(t - s) \text{ for } 0 \leq s \leq t \leq T,$$

$(H2)$ *there exists a nonnegative real number c such that*

$$\|A(t, x)\| \leq c(1 + \|x\|) \text{ for } t \in [0, T],\ x \in H.$$

Assume further that $A(t)$ satisfies

A-1. *$(t, x) \to A(t)x$ is a Caratheodory mapping, that is, $t \mapsto A(t)x$ is Lebesgue-measurable on $[0, T]$ for each fixed $x \in H$, and $x \mapsto A(t)x$ is continuous on H for each fixed $t \in [0, T]$,*

A-2. *$\langle A(t)x, x\rangle \geq \gamma ||x||^2$, for all $(t, x) \in [0, T] \times H$, for some $\gamma > 0$.*

Let $n \in \mathbb{N}$ and $\varphi_n : H \to \mathbb{R}^+$ be a C^1, convex, Lipschitz function and such that $\nabla\varphi_n$ is locally Lipschitz, and let φ_∞ be a nonnegative l.s.c proper function defined on H with $\varphi_n(x) \leq \varphi_\infty(x)$, $\forall x \in H$. For each $n \in \mathbb{N}$, let u^n be the unique $W_H^{2,\infty}([0, T])$ solution to the problem

$$\begin{cases} 0 = \ddot{u}^n(t) + A(t)\dot{u}^n(t) + \nabla\varphi_n(u^n(t)), t \in [0, T] \\ u^n(0) = u_0^n,\ \dot{u}^n(0) = \dot{u}_0^n \end{cases}$$

Assume that

(i) *φ_n epiconverges to φ_∞,*

(ii) $u^n(0) \to u_0^\infty \in \operatorname{dom}\varphi_\infty$ *and* $\lim_n \varphi_n(u^n(0)) = \varphi_\infty(u_0^\infty)$,

(iii) $\sup_{v\in \overline{B}_{L_H^\infty([0,T])}} \int_0^T \varphi_\infty(v(t))dt < +\infty$, *where* $\overline{B}_{L_H^\infty([0,T])}$ *is the closed unit ball in* $L_H^\infty([0,T])$.

(a) *Then up to extracted subsequences,* (u^n) *converges uniformly to a* $W_{BV}^{1,1}([0,T],\mathbb{R}^d)$*-function* u^∞ *with* $u^\infty(0) \in \operatorname{dom}\varphi_\infty$, *and* $(\dot u^n)$ *pointwisely converges to a BV function* v^∞ *with* $v^\infty = \dot u^\infty$, *and* $(\ddot u^n)$ *biting converges to a function* $\zeta^\infty \in L^1_{\mathbb{R}^d}([0,T])$ *so that the limit function* u^∞, $\dot u^\infty$ *and the biting limit* ζ^∞ *satisfy the variational inclusion*

$$-A(.)\dot u^\infty - \zeta^\infty \in \partial I_{\varphi_\infty}(u^\infty)$$

where $\partial I_{\varphi_\infty}$ *denotes the subdifferential of the convex lower semicontinuous integral functional* I_{φ_∞} *defined on* $L^\infty_{\mathbb{R}^d}([0,T])$

$$I_{\varphi_\infty}(u) := \int_0^T \varphi_\infty(u(t))\,dt,\ \ \forall u \in L^\infty_{\mathbb{R}^d}([0,T]).$$

(b) $(\ddot u^n)$ *weakly converges to a vector measure* $m \in \mathscr{M}_H^b([0,T])$ *so that the limit functions* $u^\infty(.)$ *and the limit measure* m *satisfy the following variational inequality:*

$$\int_0^T \varphi_\infty(v(t))\,dt \geq \int_0^T \varphi_\infty(u^\infty(t))\,dt + \int_0^T \langle -A(t)\dot u^\infty(t), v(t) - u^\infty(t)\rangle\,dt$$
$$+ \langle -m, v - u^\infty\rangle_{(\mathscr{M}^b_{\mathbb{R}^d}([0,T]),\mathscr{C}_E([0,T]))}.$$

(c) *Furthermore* $\lim_n \int_0^T \varphi_n(u^n(t))dt = \int_0^T \varphi_\infty(u^\infty(t))dt$. *Subsequently the energy estimate*

$$\varphi_\infty(u^\infty(t)) + \frac{1}{2}||\dot u^\infty(t)||^2 = \varphi_\infty(u_0^\infty) + \frac{1}{2}||\dot u_0^\infty||^2 + \int_0^t \langle -A(s)\dot u^\infty(s), u^\infty(s)\rangle ds$$

holds a.e.

(d) *There is a filter* $\mathscr{U}$ *finer than the Fréchet filter* $l \in L^\infty_{\mathbb{R}^d}([0,T])'$ *such that*

$$\mathscr{U} - \lim_n [-A(.)\dot u^n - \ddot u^n] = l \in L^\infty_{\mathbb{R}^d}([0,T])'_{\text{weak}}$$

where $L^\infty_{\mathbb{R}^d}([0,T])'_{\text{weak}}$ *is the second dual of* $L^1_{\mathbb{R}^d}([0,T])$ *endowed with the topology* $\sigma(L^\infty_{\mathbb{R}^d}([0,T])', L^\infty_{\mathbb{R}^d}([0,T]))$, *and* $\mathbf{n} \in \mathscr{C}_{\mathbb{R}^d}([0,T])'_{\text{weak}}$ *such that*

$$\lim_n [-A(.)\dot u^n - \ddot u^n] = \mathbf{n} \in \mathscr{C}_{\mathbb{R}^d}([0,T])'_{\text{weak}}$$

where $\mathscr{C}_{\mathbb{R}^d}([0,T])'_{\text{weak}}$ *denotes the space* $\mathscr{C}_{\mathbb{R}^d}([0,T])'$ *endowed with the weak topology* $\sigma(\mathscr{C}_{\mathbb{R}^d}([0,T])', \mathscr{C}_{\mathbb{R}^d}([0,T]))$. *Let* l_a *be the density of the absolutely continuous part* l_a *of* l *in the decomposition* $l = l_a + l_s$ *in absolutely continuous part* l_a *and singular part* l_s. *Then*

$$l_a(f) = \int_0^T \langle f(t), -A(t)\dot{u}^\infty(t) - \zeta^\infty(t)\rangle dt$$

for all $f \in L^\infty_{\mathbb{R}^d}([0,T])$ *so that*

$$I^*_{\varphi_\infty}(l) = I_{\varphi^*_\infty}(-A(.)\dot{u}^\infty - \zeta^\infty) + \delta^*(l_s, \operatorname{dom} I_{\varphi_\infty})$$

where φ^*_∞ *is the conjugate of* φ_∞, $I_{\varphi^*_\infty}$ *the integral functional defined on* $L^1_{\mathbb{R}^d}([0,T])$ *associated with* φ^*_∞, $I^*_{\varphi_\infty}$ *the conjugate of the integral functional* I_{φ_∞}, $\operatorname{dom} I_{\varphi_\infty} := \{u \in L^\infty_{\mathbb{R}^d}([0,1]) : I_{\varphi_\infty}(u) < \infty\}$, *and*

$$\langle \mathbf{n}, f\rangle = \int_0^T \langle -A(t)\dot{u}^\infty(t) - \zeta^\infty(t), f(t)\rangle dt + \langle \mathbf{n}_s, f\rangle, \quad \forall f \in \mathscr{C}_{\mathbb{R}^d}([0,T]).$$

with $\langle \mathbf{n}_s, f\rangle = l_s(f)$, $\forall f \in \mathscr{C}_{\mathbb{R}^d}([0,T])$. *Further* $\mathbf{n}$ *belongs to the subdifferential* $\partial J_{\varphi_\infty}(u^\infty)$ *of the convex lower semicontinuous integral functional* J_{φ_∞} *defined on* $\mathscr{C}_{\mathbb{R}^d}([0,T])$

$$J_{\varphi_\infty}(u) := \int_0^T \varphi_\infty(u(t))\, dt, \ \forall u \in \mathscr{C}_{\mathbb{R}^d}([0,T]).$$

Consequently the density $-A(.)\dot{u}^\infty - \zeta^\infty$ *of the absolutely continuous part* $\mathbf{n}_a$

$$\mathbf{n}_a(f) := \int_0^T \langle -A(t)\dot{u}^\infty(t) - \zeta^\infty(t), f(t)\rangle dt, \quad \forall f \in \mathscr{C}_{\mathbb{R}^d}([0,T])$$

satisfies the inclusion

$$-A(t)\dot{u}^\infty(t) - \zeta^\infty(t) \in \partial\varphi_\infty(u^\infty(t)), \quad \text{a.e.}$$

and for any nonnegative measure θ *on* $[0,T]$ *with respect to which* $\mathbf{n}_s$ *is absolutely continuous*

$$\int_0^T r_{\varphi^*_\infty}\left(\frac{d\mathbf{n}_s}{d\theta}(t)\right) d\theta(t) = \int_0^T \left\langle u^\infty(t), \frac{d\mathbf{n}_s}{d\theta}(t)\right\rangle d\theta(t)$$

where $r_{\varphi^*_\infty}$ *denotes the recession function of* φ^*_∞.

Proof The proof is long and based on the existence and uniqueness of $W_H^{2,\infty}([0,T])$ solution to the approximating equation (cf. Corollary 3.2)

$$\begin{cases} 0 = \ddot{u}^n(t) + A(t)\dot{u}^n(t) + \nabla\varphi_n(u^n(t)), t \in [0,T] \\ u^n(0) = u_0^n,\ \dot{u}^n(0) = \dot{u}_0^n \end{cases}$$

and the techniques developed in [22, 23, 27]. Nevertheless we will produce the proof with full details, since the techniques employed can be applied to further related results.

Step 1. Multiplying scalarly the equation

$$-A(t)\dot{u}^n(t) - \ddot{u}^n(t) = \nabla\varphi_n(u^n(t))$$

by $\dot{u}^n(t)$ and applying the chain rule theorem [42, Theorem 2] yields

$$-\langle \dot{u}^n(t), A(t)\dot{u}^n(t)\rangle - \langle \dot{u}^n(t), \ddot{u}^n(t)\rangle = \frac{d}{dt}[\varphi_n(u_n(t))],$$

that is,

$$-\langle \dot{u}^n(t), A(t)\dot{u}^n(t)\rangle = \frac{d}{dt}\left[\varphi_n(u^n(t)) + \frac{1}{2}||\dot{u}^n(t)||^2\right].$$

By integrating on $[0,t]$ this equality and using the condition (ii), we get

$$\begin{aligned} \varphi_n(u^n(t)) + \frac{1}{2}||\dot{u}^n(t)||^2 &= \varphi_n(u^n(0)) + \frac{1}{2}||\dot{u}^n(0)||^2 - \int_0^t \langle \dot{u}^n(s), A(s)\dot{u}^n(s)\rangle ds \\ &\leq \varphi_n(u^n(0)) + \frac{1}{2}||\dot{u}^n(0)||^2 + \gamma \int_0^t ||\dot{u}^n(s||^2 ds. \end{aligned}$$

Then, from our assumption, $\varphi_n(u^n(0)) \leq$ positive constant $< +\infty$ and $\frac{1}{2}||\dot{u}^n(0)||^2 \leq$ positive constant $< +\infty$ so that

$$\varphi_n(u^n(t)) + \frac{1}{2}||\dot{u}^n(t)||^2 \leq p + \gamma \int_0^t ||\dot{u}^n(s||^2 ds,\ t \in [0,T]$$

where p is a generic positive constant. So by the preceding estimate and the Gronwall inequality [21, Lemma 3.1] , it is immediate that

$$\sup_{n\geq 1}\sup_{t\in[0,T]} ||\dot{u}^n(t)|| < +\infty \quad \text{and} \quad \sup_{n\geq 1}\sup_{t\in[0,T]} \varphi_n(u^n(t)) < +\infty. \tag{1}$$

Step 2. Estimation of $||\ddot{u}^n(.)||$. For simplicity, let us set $z^n(t) = -A(t)\dot{u}^n(t) - \ddot{u}^n(t), \forall t \in [0,T]$. As

$$z^n(t) := -A(t)\dot{u}^n(t) - \ddot{u}^n(t) = \nabla\varphi_n(u^n(t))$$

by the subdifferential inequality for convex lower semicontinuous functions, we have

$$\varphi_n(x) \geq \varphi_n(u^n(t)) + \langle x - u^n(t), z^n(t)\rangle$$

for all $x \in \mathbb{R}^d$. Now let $v \in \overline{B}_{L^\infty_{\mathbb{R}^d}([0,T])}$, the closed unit ball of $L^\infty_{\mathbb{R}^d}[0, T]$). By taking $x = v(t)$ in the preceding inequality, we get

$$\varphi_n(v(t)) \geq \varphi_n(u^n(t)) + \langle v(t) - u^n(t), z^n(t)\rangle.$$

Integrating the preceding inequality gives

$$\int_0^T \langle v(t) - u^n(t), z^n(t)\rangle dt \leq \int_0^T \varphi_n(v(t))dt - \int_0^T \varphi_n(u^n(t))dt.$$

Whence follows

$$\begin{aligned}&\int_0^T \langle v(t), z^n(t)\rangle dt \\ &\qquad \leq \int_0^T \varphi_n(v(t))dt - \int_0^T \varphi_n(u^n(t))dt + \int_0^1 \langle u^n(t), z^n(t)\rangle dt. \qquad (2)\end{aligned}$$

We compute the last integral in the preceding inequality. By integration and taking account of (1), we have

$$\begin{aligned}&\int_0^T \langle u^n(t), z^n(t)\rangle dt \\ &= \int_0^T \langle u^n(t), -A(t)\dot{u}^n(t) - \ddot{u}^n(t)\rangle dt \\ &= -\left[\langle u^n(t), \dot{u}^n(t)\rangle\right]_0^T + \int_0^T \langle \dot{u}^n(t), \dot{u}^n(t)\rangle dt - \int_0^T \langle u^n(t), A(t)\dot{u}^n(t)\rangle dt \\ &= -\langle u^n(T), \dot{u}^n(T)\rangle + \langle u^n(0), \dot{u}^n(0)\rangle \\ &\qquad + \int_0^T ||\dot{u}^n(t)||^2 dt - \int_0^T \langle u^n(t), A(t)\dot{u}^n(t)\rangle dt. \qquad (3)\end{aligned}$$

As $||A(t)\dot{u}^n(t)|| \leq c(1 + ||\dot{u}_n(t)||)$ by (H_2), so that by (1) it is immediate that $\int_0^T \langle u^n(t), A(t)\dot{u}^n(t)\rangle dt$ is uniformly bounded so that by (1), (2), and (3), we get

$$\begin{aligned}&\int_0^T \langle v(t), z^n(t)\rangle dt \leq \int_0^T \varphi_n(v(t))dt + L \\ &\qquad \leq \sup_{v \in \overline{B}_{L^\infty_{\mathbb{R}^d}([0,T])}} \int_0^T \varphi_\infty(v(t))dt + L < \infty \qquad (4)\end{aligned}$$

for all $v \in \overline{B}_{L^\infty_{\mathbb{R}^d}([0,T])}$. Here L is a generic positive constant independent of $n \in \mathbb{N}$. By (4), we conclude that $(z^n = -A(.)\dot{u}^n - \ddot{u}^n)$ is bounded in $L^1_{\mathbb{R}^d}([0,T])$, and then so is $(\ddot{u}^n)$. It turns out that the sequence $(\dot{u}^n)$ of absolutely continuous functions is uniformly bounded by (1) and bounded in variation and by Helly's theorem; we may assume that $(\dot{u}^n)$ pointwisely converges to a BV function $v^\infty : [0,T] \to \mathbb{R}^d$ and the sequence (u^n) converges uniformly to an absolutely continuous function u^∞ with $\dot{u}^\infty = v^\infty$ a.e. At this point, it is clear that $A(t)\dot{u}^n(t) \to A(t)v^\infty(t)$ so that $A(t)\dot{u}^n(t) \to A(t)\dot{u}^\infty(t)$ a.e. and $A(.)\dot{u}^n(.)$ converges in $L^1_{\mathbb{R}^d}([0,T])$ to $A(.)\dot{u}^\infty(.)$, using (1) and the dominated convergence theorem.

Step 3. Young measure limit and biting limit of $\ddot{u}_n$. As $(\ddot{u}_n)$ is bounded in $L^1_{\mathbb{R}^d}([0,T])$, we may assume that $(\ddot{u}^n)$ stably converges to a Young measure $\nu \in \mathscr{Y}([0,T]); \mathbb{R}^d)$ with $\mathrm{bar}(\nu) : t \mapsto \mathrm{bar}(\nu_t) \in L^1_{\mathbb{R}^d}([0,T])$ (here $\mathrm{bar}(\nu_t)$ denotes the barycenter of ν_t). Further, we may assume that $(\ddot{u}^n)$ biting converges to a function $\zeta^\infty : t \mapsto \mathrm{bar}(\nu_t)$, that is, there exists a decreasing sequence of Lebesgue-measurable sets (B_p) with $\lim_p \lambda(B_p) = 0$ such that the restriction of $(\ddot{u}_n)$ on each B_p^c converges weakly in $L^1_{\mathbb{R}^d}([0,T])$ to ζ^∞. Noting that $(A(.)\dot{u}^n)$ converges in $L^1_{\mathbb{R}^d}([0,T])$ to $A(.)\dot{u}^\infty$. It follows that the restriction of $z^n = -A(.)\dot{u}^n - \ddot{u}^n$ to each B_p^c weakly converges in $L^1_{\mathbb{R}^d}([0,T])$ to $z^\infty := -A(.)\dot{u}^\infty - \zeta^\infty$, because $(-A(.)\dot{u}^n)$ converges in $L^1_{\mathbb{R}^d}([0,T])$ to $A(.)\dot{u}^\infty$ and $(\ddot{u}^n)$ biting converges to $\zeta^\infty \in L^1_{\mathbb{R}^d}([0,T])$. It follows that

$$\lim_n \int_B \langle -A(.)\dot{u}^n - \ddot{u}^n, w(t) - u^n(t)\rangle = \int_B \langle -A(.)\dot{u}^\infty - \mathrm{bar}(\nu_t), w(t) - u(t)\rangle dt \tag{5}$$

for every $B \in B_p^c \cap \mathscr{L}([0,T])$ and for every $w \in L^\infty_{\mathbb{R}^d}([0,T])$. Indeed, we note that $(w(t) - u^n(t))$ is a bounded sequence in $L^\infty_{\mathbb{R}^d}([0,T])$ which pointwisely converges to $w(t) - u^\infty(t)$, so it converges uniformly on every uniformly integrable subset of $L^1_{\mathbb{R}^d}([0,T])$ by virtue of a Grothendieck Lemma [33], recalling here that the restriction of $-A(.)\dot{u}^n - \ddot{u}^n$ on each B_p^c is uniformly integrable. Now, since φ_n lower epiconverges to φ_∞, for every Lebesgue-measurable set A in $[0,T]$, by virtue of [23, Corollary 4.7], we have

$$+\infty > \liminf_n \int_A \varphi_n(u^n(t))dt \geq \int_A \varphi_\infty(u^\infty(t))dt. \tag{6}$$

Combining (1), (2), (3), (4), (5), and (6) and using the subdifferential inequality

$$\varphi_n(w(t)) \geq \varphi_n(u^n(t)) + \langle -A(.)\dot{u}^n - \ddot{u}^n(t), w(t) - u^n(t)\rangle,$$

we get

$$\int_B \varphi_\infty(w(t))\,dt \geq \int_B \varphi_\infty(u^\infty(t))\,dt + \int_B \langle -A(.)\dot{u}^\infty - \mathrm{bar}(\nu_t), w(t) - u^\infty(t)\rangle\,dt.$$

This shows that $t \mapsto -A(.)\dot{u}^\infty - \text{bar}(\nu_t)$ is a subgradient at the point u^∞ of the convex integral functional I_{φ_∞} restricted to $L^\infty_{\mathbb{R}^d}(B_p^c)$, consequently,

$$-A(.)\dot{u}^\infty - \text{bar}(\nu_t) \in \partial\varphi_\infty(u^\infty(t)), \text{ a.e. on } B_p^c.$$

As this inclusion is true on each B_p^c and $B_p^c \uparrow [0, T]$, we conclude that

$$-A(.)\dot{u}^\infty - \text{bar}(\nu_t) \in \partial\varphi_\infty(u^\infty(t)), \text{ a.e. on } [0, T].$$

Step 4. Measure limit in $\mathcal{M}^b_{\mathbb{R}^d}([0, T])$ *of* $\ddot{u}^n$. As $(\ddot{u}_n)$ is bounded in $L^1_{\mathbb{R}^d}([0, T])$, we may assume that $(\ddot{u}^n)$ weakly converges to a vector measure $m \in \mathcal{M}^b_{\mathbb{R}^d}([0, T])$ so that the limit functions $u^\infty(.)$ and the limit measure m satisfy the following variational inequality:

$$\int_0^T \varphi_\infty(v(t))\,dt \geq \int_0^T \varphi_\infty(u^\infty(t))\,dt + \int_0^T \langle -A(t)\dot{u}^\infty(t), v(t) - u^\infty(t)\rangle\,dt$$
$$+ \langle -m, v - u^\infty\rangle_{(\mathcal{M}^b_E([0,T]), \mathcal{C}_{\mathbb{R}^d}([0,T]))}.$$

In other words, the vector measure $-m - A(t)\dot{u}^\infty(t)dt$ belongs to the subdifferential $\partial J_{\varphi_\infty}(u^\infty)$ of the convex functional integral J_{φ_∞} defined on $\mathcal{C}_{\mathbb{R}^d}([0, T])$ by $J_{\varphi_\infty}(v) = \int_0^T \varphi_\infty(v(t))\,dt$, $\forall v \in \mathcal{C}_{\mathbb{R}^d}([0, T])$. Indeed, let $w \in \mathcal{C}_{\mathbb{R}^d}([0, T])$. Integrating the subdifferential inequality

$$\varphi_n(w(t)) \geq \varphi_n(u^n(t)) + \langle -A(t)\dot{u}^n(t) - \ddot{u}^n(t), w(t) - u^n(t)\rangle$$

and noting that $\varphi_\infty(w(t)) \geq \varphi_n(w(t))$ gives immediately

$$\int_0^T \varphi_\infty(w(t))dt \geq \int_0^T \varphi_n(w(t))dt$$
$$\geq \int_0^T \varphi_n(u^n(t))dt + \langle -A(t)\dot{u}^n(t) - \ddot{u}^n(t), w(t) - u^n(t)\rangle dt.$$

We note that

$$\lim_n \int_0^T \langle -A(t)\dot{u}^n(t), w(t) - u^n(t)\rangle dt = \int_0^T \langle A(t)\dot{u}^\infty(t), w(t) - u^\infty(t)\rangle dt$$

because $(-A(.)\dot{u}^n)$ is uniformly integrable and converges in $L^1_H([0, T])$ to $A(.)\dot{u}^\infty$ and the sequence in $(w - u^n)$ converges uniformly to $w - u^\infty$. Whence follows

$$\int_0^T \varphi_\infty(w(t))dt \geq \int_0^T \varphi_\infty(u^\infty(t))dt + \int_0^T \langle -A(t)\dot{u}^\infty(t), w(t) - u^\infty(t)\rangle dt$$
$$+ \langle -m, w - u^\infty\rangle_{(\mathcal{M}^b_{\mathbb{R}^d}([0,T]), \mathcal{C}_{\mathbb{R}^d}([0,T]))},$$

which shows that the vector measure $-m - A(.)\dot{u}^\infty dt$ is a subgradient at the point u^∞ of the of the convex integral functional J_{φ_∞} defined on $\mathscr{C}_{\mathbb{R}^d}([0,T]))$ by $J_{\varphi_\infty}(v) := \int_0^T \varphi_\infty(v(t))dt, \forall v \in \mathscr{C}_{\mathbb{R}^d}([0,T])$.

Step 5. Claim $\lim_n \varphi_n(u^n(t)) = \varphi_\infty(u^\infty(t)) < \infty$ a.e. and $\lim_n \int_0^T \varphi_n(u^n(t))dt = \int_0^T \varphi_\infty(u^\infty(t))dt < \infty$, and subsequently, the energy estimate holds for a.e. $t \in [0,T]$:

$$\varphi_\infty(u^\infty(t)) + \frac{1}{2}||\dot{u}^\infty(t)||^2 = \varphi_\infty(u_0^\infty) + \frac{1}{2}||\dot{u}_0^\infty||^2 - \int_0^t \langle A(s)(\dot{u}^\infty(s), \dot{u}^\infty(s)\rangle ds.$$

With the above stated results and notations, applying the subdifferential inequality

$$\varphi_n(w(t)) \geq \varphi_n(u^n(t)) + \langle -A(t)\dot{u}^n(t) - \ddot{u}^n(t), w(t) - u^n(t)\rangle$$

with $w = u^\infty$, integrating on $B \in B_p^c \cap \mathscr{L}([0,T])$, and passing to the limit when n goes to ∞, gives the inequality

$$\begin{aligned}\int_B \varphi_\infty(u^\infty(t))dt &\geq \liminf_n \int_B \varphi_n(u^n(t))dt \\ &\geq \int_B \varphi_\infty(u^\infty(t))dt \geq \limsup_n \int_B \varphi_n(u^n(t))dt\end{aligned}$$

so that

$$\lim_n \int_B \varphi_n(u^n(t))dt = \int_B \varphi_\infty(u^\infty(t))dt \tag{7}$$

on $B \in B_p^c \cap \mathscr{L}([0,T])$. Now, from the chain rule theorem given in Step 1, recall that

$$-\langle \dot{u}^n(t), A(t)\dot{u}^n(t)\rangle - \langle \dot{u}^n(t), \ddot{u}^n(t)\rangle = \frac{d}{dt}[\varphi_n(u_n(t))],$$

that is,

$$\langle \dot{u}^n(t), z^n(t)\rangle = \frac{d}{dt}[\varphi_n(u_n(t))].$$

By the estimate (1) and the boundedness in $L^1_{\mathbb{R}^d}([0,T])$ of (z^n), it is immediate that $(\frac{d}{dt}[\varphi_n(u_n(t))])$ is bounded in $L^1_{\mathbb{R}}([0,T])$ so that $(\varphi_n(u_n(.))$ is bounded in variation. By Helly's theorem, we may assume that $(\varphi_n(u_n(.))$ pointwisely converges to a BV function ψ. By (1), $(\varphi_n(u_n(.))$ converges in $L^1_{\mathbb{R}}([0,T])$ to ψ. In particular, for every $k \in L^\infty_{\mathbb{R}^+}([0,T])$, we have

$$\lim_{n\to\infty} \int_0^T k(t)\varphi_n(u_n(t))dt = \int_0^T k(t)\psi(t)dt. \tag{8}$$

Combining with (7) and (8) yields

$$\int_B \psi(t)\,dt = \lim_{n\to\infty} \int_B \varphi_n(u^n(t))\,dt = \int_B \varphi_\infty(u^\infty(t))\,dt$$

for all $\in B_p^c \cap \mathscr{L}([0,T])$. As this inclusion is true on each B_p^c and $B_p^c \uparrow [0,T]$, we conclude that

$$\psi(t) = \lim_n \varphi_n(u_n(t)) = \varphi_\infty(u^\infty(t)) \text{ a.e.}$$

Subsequently, using (iii), the passage to the limit when n goes to ∞ in the equation

$$\varphi_n(u^n(t)) + \frac{1}{2}||\dot{u}^n(t)||^2 = \varphi_n(u^n(0)) + \frac{1}{2}||\dot{u}^n(0)||^2 - \int_0^t \langle A(s)\dot{u}^n(s), \dot{u}^n(s)\rangle ds$$

yields for a.e. $t \in [0,T]$

$$\varphi_\infty(u^\infty(t)) + \frac{1}{2}||\dot{u}^\infty(t)||^2 = \varphi_\infty(u_0^\infty) + \frac{1}{2}||\dot{u}_0^\infty)||^2 - \int_0^t \langle A(s)\dot{u}^\infty(s), \dot{u}^\infty(s)\rangle ds.$$

Step 6. Localization of further limits and final step.
As $(z^n = -A(.)\dot{u}^n - \ddot{u}^n)$ is bounded in $L^1_{\mathbb{R}^d}([0,T])$ in view of Step 3, it is relatively compact in the second dual $L^\infty_{\mathbb{R}^d}([0,T])'$ of $L^1_{\mathbb{R}^d}([0,T])$ endowed with the weak topology $\sigma(L^\infty_{\mathbb{R}^d}([0,T])', L^\infty_{\mathbb{R}^d}([0,T]))$. Furthermore, (z^n) can be viewed as a bounded sequence in $\mathscr{C}_{\mathbb{R}^d}([0,T])'$. Hence there is a filter $\mathscr{U}$ finer than the Fréchet filter $l \in L^\infty_{\mathbb{R}^d}([0,T])'$ and $\mathbf{n} \in \mathscr{C}_{\mathbb{R}^d}([0,T])'$ such that

$$\mathscr{U} - \lim_n z^n = l \in L^\infty_{\mathbb{R}^d}([0,T])'_{\text{weak}} \tag{9}$$

and

$$\lim_n z^n = \mathbf{n} \in \mathscr{C}_{\mathbb{R}^d}([0,T])'_{\text{weak}} \tag{10}$$

where $L^\infty_{\mathbb{R}^d}([0,T])'_{\text{weak}}$ is the second dual of $L^1_{\mathbb{R}^d}([0,T])$ endowed with the topology $\sigma(L^\infty_{\mathbb{R}^d}([0,T])', L^\infty_{\mathbb{R}^d}([0,T]))$ and $\mathscr{C}_{\mathbb{R}^d}([0,T])'_{\text{weak}}$ denotes the space $\mathscr{C}_{\mathbb{R}^d}([0,T])'$ endowed with the weak topology $\sigma(\mathscr{C}_{\mathbb{R}^d}([0,T])', \mathscr{C}_{\mathbb{R}^d}([0,T]))$, because $\mathscr{C}_{\mathbb{R}^d}([0,T])$ is a separable Banach space for the norm sup, so that we may assume by extracting subsequences that (z^n) weakly converges to $\mathbf{n} \in \mathscr{C}_{\mathbb{R}^d}([0,T])'$. Let l_a be the density of the absolutely continuous part l_a of l in the decomposition $l = l_a + l_s$ in absolutely continuous part l_a and singular part l_s, in the sense there is a decreasing sequence (A_n) of Lebesgue-measurable sets in $[0,T]$ with $A_n \downarrow \emptyset$ such that $l_s(f) = l_s(1_{A_n} f)$ for all $h \in L^\infty_{\mathbb{R}^d}([0,T])$ and for all $n \geq 1$. As $(z^n = -A(.)\dot{u}^n - \ddot{u}^n)$ biting converges to $z^\infty = -A(.)\dot{u}^\infty - \zeta^\infty$ in Step 4, it is already known [22] that

$$l_a(f) = \int_0^T \langle f(t), -A(t)\dot{u}^\infty(t) - \zeta^\infty(t)\rangle dt$$

for all $f \in L^\infty_{\mathbb{R}^d}([0,T])$, shortly $z^\infty = -A(t)\dot{u}^\infty(t) - \zeta^\infty(t)$ coincides a.e. with the density of the absolutely continuous part l_a. By [19, 46], we have

$$I^*_{\varphi_\infty}(l) = I_{\varphi^*_\infty}(-A(.)\dot{u}^\infty - \zeta^\infty) + \delta^*(l_s, \mathrm{dom}\, I_{\varphi_\infty})$$

where φ^*_∞ is the conjugate of φ_∞, $I_{\varphi^*_\infty}$ is the integral functional defined on $L^1_{\mathbb{R}^d}([0,T])$ associated with φ^*_∞, $I^*_{\varphi_\infty}$ is the conjugate of the integral functional I_{φ_∞}, and

$$\mathrm{dom}\, I_{\varphi_\infty} := \{u \in L^\infty_{\mathbb{R}^d}([0,T]) : I_{\varphi_\infty}(u) < \infty\}.$$

Using the inclusion

$$z^\infty = -A(.)\dot{u}^\infty - \zeta^\infty \in \partial I_{\varphi_\infty}(u^\infty),$$

that is,

$$I_{\varphi^*_\infty}(-A(.)\dot{u}^\infty - \zeta^\infty) = \langle -A(.)\dot{u}^\infty - \zeta^\infty, u^\infty\rangle - I_{\varphi_\infty}(u^\infty),$$

we see that

$$I^*_{\varphi_\infty}(l) = \langle -A(.)\dot{u}^\infty - \zeta^\infty, u^\infty\rangle - I_{\varphi_\infty}(u^\infty) + \delta^*(l_s, \mathrm{dom}\, I_{\varphi_\infty}).$$

Coming back to $z^n(t) = \nabla\varphi_n(u^n(t))$, we have

$$\varphi_n(x) \geq \varphi_n(u^n(t)) + \langle x - u^n(t), z^n(t)\rangle$$

for all $x \in \mathbb{R}^d$. Substituting x by $h(t)$ in this inequality, where $h \in \mathscr{C}_{\mathbb{R}^d}([0,T])$, and integrating, we get

$$\int_0^T \varphi_n(h(t))\,dt \geq \int_0^T \varphi_n(u^n(t))\,dt + \int_0^T \langle h(t) - u^n(t), z^n(t)\rangle\,dt.$$

Arguing as in Step 4 by passing to the limit in the preceding inequality, involving the epiliminf property for integral functionals (cf. (6)), it is easy to see that

$$\int_0^T \varphi_\infty(h(t))\,dt \geq \int_0^T \varphi_\infty(u^\infty(t))\,dt + \langle h - u^\infty, \mathbf{n}\rangle.$$

Whence $\mathbf{n}$ belongs to the subdifferential $\partial J_{\varphi_\infty}(u^\infty)$ of the convex lower semicontinuous integral functional J_{φ_∞} defined on $\mathscr{C}_{\mathbb{R}^d}([0,T])$ by

$$J_{\varphi_\infty}(u) := \int_0^T \varphi_\infty(u(t))\,dt, \ \forall u \in \mathscr{C}_{\mathbb{R}^d}([0,T]).$$

Now let $B : \mathscr{C}_{\mathbb{R}^d}([0,T]) \to L^\infty_{\mathbb{R}^d}([0,T])$ be the continuous injection, and let $B^* : L^\infty_{\mathbb{R}^d}([0,T])' \to \mathscr{C}_{\mathbb{R}^d}([0,T])'$ be the adjoint of B given by

$$\langle B^* l, f\rangle = \langle l, Bf\rangle = \langle l, f\rangle, \quad \forall l \in L^\infty_{\mathbb{R}^d}([0,T])', \quad \forall f \in \mathscr{C}_{\mathbb{R}^d}([0,T]).$$

Then we have $B^*l = B^*l_a + B^*l_s$, $l \in L^\infty_{\mathbb{R}^d}([0,T])'$ being the limit of z_n under the filter $\mathscr{U}$ given in Sect. 4 and $l = l_a + l_s$ being the decomposition of l in absolutely continuous part l_a and singular part l_s. It follows that

$$\langle B^*l, f\rangle = \langle B^*l_a, f\rangle + \langle B^*l_s, f\rangle = \langle l_a, f\rangle + \langle l_s, f\rangle$$

for all $f \in \mathscr{C}_{\mathbb{R}^d}([0,T])$. But it is already seen that

$$\begin{aligned}\langle l_a, f\rangle &= \langle -A(.)\dot u^\infty - \zeta^\infty, f\rangle \\ &= \int_0^T \langle -A(.)\dot u^\infty(t) - \zeta^\infty(t), f(t)\rangle dt, \quad \forall f \in L^\infty_{\mathbb{R}^d}([0,T])\end{aligned}$$

so that the measure B^*l_a is absolutely continuous

$$\langle B^*l_a, h\rangle = \int_0^T \langle -A(.)\dot u^\infty(t) - \zeta^\infty(t), f(t)\rangle dt, \quad \forall f \in \mathscr{C}_{\mathbb{R}^d}([0,T])$$

and its density $-A(.)\dot u^\infty - \zeta^\infty$ satisfies the inclusion

$$-A(t)\dot u^\infty(t) - \zeta^\infty(t) \in \partial\varphi_\infty(u^\infty(t)), \quad \text{a.e.}$$

and the singular part B^*l_s satisfies the equation

$$\langle B^*l_s, f\rangle = \langle l_s, h\rangle, \quad \forall f \in \mathscr{C}_{\mathbb{R}^d}([0,T]).$$

As $B^*l = \mathbf{n}$, using (9) and (10), it turns out that $\mathbf{n}$ is the sum of the absolutely continuous measure $\mathbf{n}_a$ with

$$\langle \mathbf{n}_a, f\rangle = \int_0^T \langle -A(t)\dot u^\infty(t) - \zeta^\infty(t), f(t)\rangle dt, \quad \forall f \in \mathscr{C}_{\mathbb{R}^d}([0,T])$$

and the singular part $\mathbf{n}_s$ given by

$$\langle \mathbf{n}_s, f\rangle = \langle l_s, f\rangle, \quad \forall f \in \mathscr{C}_{\mathbb{R}^d}([0,T]).$$

which satisfies the property: for any nonnegative measure θ on $[0, T]$ with respect to which $\mathbf{n}_s$ is absolutely continuous

$$\int_0^T r_{\varphi_\infty^*}\left(\frac{d\mathbf{n}_s}{d\theta}(t)\right) d\theta(t) = \int_0^T \left\langle u^\infty(t), \frac{d\mathbf{n}_s}{d\theta}(t)\right\rangle d\theta(t)$$

where $r_{\varphi_\infty^*}$ denotes the recession function of φ_∞^*. Indeed, as $\mathbf{n}$ belongs to $\partial J_{\varphi_\infty}(u^\infty)$ by applying [46, Theorem 5], we have

$$J_{\varphi_\infty}^*(n) = I_{\varphi_\infty^*}\left(\frac{d\mathbf{n}_a}{dt}\right) + \int_0^T r_{\varphi_\infty^*}\left(\frac{d\mathbf{n}_s}{d\theta}(t)\right) d\theta(t), \tag{11}$$

with

$$I_{\varphi_\infty^*}(v) := \int_0^T \varphi_\infty^*(v(t))dt, \forall v \in L^1_{\mathbb{R}^d}([0, T]).$$

Recall that

$$\frac{d\mathbf{n}_a}{dt} = -A(.)\dot{u}^\infty - \zeta^\infty \in \partial I_{\varphi_\infty}(u^\infty),$$

that is,

$$I_{\varphi_\infty^*}\left(\frac{d\mathbf{n}_a}{dt}\right) = \langle -A(.)\dot{u}^\infty - \zeta^\infty, u^\infty\rangle_{\langle L^1_{\mathbb{R}^d}([0,T]), L^\infty_{\mathbb{R}^d}([0,T])\rangle} - I_{\varphi_\infty}(u^\infty). \tag{12}$$

From (12), we deduce

$$\begin{aligned} J_{\varphi_\infty}^*(n) =& \langle u^\infty, \mathbf{n}\rangle_{\langle \mathscr{C}_{\mathbb{R}^d}([0,T]), \mathscr{C}_{\mathbb{R}^d}([0,T])'\rangle} - J_{\varphi_\infty}(u^\infty) \\ =& \langle u^\infty, \mathbf{n}\rangle_{\langle \mathscr{C}_{\mathbb{R}^d}([0,T]), \mathscr{C}_{\mathbb{R}^d}([0,T])'\rangle} - I_{\varphi_\infty}(u^\infty) \\ =& \int_0^T \langle u^\infty(t), -A(.)\dot{u}^\infty - \zeta^\infty(t)\rangle dt \\ &+ \int_0^T \left\langle u^\infty(t), \frac{d\mathbf{n}_s}{d\theta}(t)\right\rangle d\theta(t) - I_{\varphi_\infty}(u^\infty) \\ =& I_{\varphi_\infty^*}\left(\frac{d\mathbf{n}_a}{dt}\right) + \int_0^T \left\langle u^\infty(t), \frac{d\mathbf{n}_s}{d\theta}(t)\right\rangle d\theta(t)). \end{aligned}$$

Coming back to (11), we get the equality

$$\int_0^T r_{\varphi_\infty^*}\left(\frac{d\mathbf{n}_s}{d\theta}(t)\right) d\theta(t) = \int_0^T \left\langle u^\infty(t), \frac{d\mathbf{n}_s}{d\theta}(t)\right\rangle d\theta(t)).$$

The proof is complete.

Comments Some comments are in order. In Proposition 3.1, using the existence and uniqueness of $W_H^{2,\infty}(]0,T])$ of the approximating second-order equation

$$\begin{cases} 0 = \ddot{u}^n(t) + A(t)\dot{u}^n(t) + \nabla\varphi_n(u^n(t)), t \in [0,T] \\ u^n(0) = u_0^n,\ \dot{u}^n(0) = \dot{u}_0^n, \end{cases}$$

we state the existence of a generalized solution u^∞ to the second-order evolution inclusion

$$\begin{cases} 0 \in \ddot{u}(t) + A(t)\dot{u}(t) + \partial\varphi_\infty(u(t)), t \in [0,T] \\ u(0) = u_0 \in \text{dom}\ \varphi_\infty,\ \dot{u}(0) = \dot{u}_0 \end{cases}$$

via an epiconvergence approach involving the structure of bounded sequences in $L_H^1([0,T]$ space [22] and describe various properties of such a generalized solution. In particular, we show that such a generalized solution u^∞ is $W_{BV}^{1,1}([0,T])$ and satisfies the energy conservation and there exists a Young measure ν_t with barycenter $\text{bar}(\nu_t) \in L_H^1([0,T])$ such that $-A(t)\dot{u}^\infty(t) - \text{bar}(\nu_t) \in \partial\varphi_\infty(u\infty(t))$ a.e. In this vein, compare with Attouch et al. [4, 27], Paoli [43], and Schatzman [48].

Now we deal at first with $W_{BV}^{1,1}([0,T],H)$ solution for a second-order evolution problem.

Theorem 3.3 *Let for every $t \in [0,T]$, $A(t) : D(A(t)) \subset H \to 2^H$ be a maximal monotone operator with $D(A(t))$ ball compact for every $t \in [0,T]$ satisfying*

$(H1)$ *there exists a function $r : [0,T] \to [0,+\infty[$ which is continuous on $[0,T[$ and nondecreasing with $r(T) < +\infty$ such that*

$$\text{dis}(A(t), A(s)) \le dr(]s,t]) = r(t) - r(s) \ \ for\ \ 0 \le s \le t \le T$$

$(H2)$ *there exists a nonnegative real number c such that*

$$\|A^0(t,x)\| \le c(1 + \|x\|) \ \ for\ \ t \in [0,T],\ x \in D(A(t))$$

Let $f : [0,T] \times H \times H \to H$ be such that for every $x, y \in H \times H$ the mapping $f(.,x,y)$ is Borel-measurable on $[0,T]$ and for every $t \in [0,T]$, $f(t,.,.)$ is continuous on $H \times H$ and satisfying

(i) $||f(t,x,y)|| \le M(1 + ||x||),\ \forall t,x,y \in [0,T] \times H \times H.$
(ii) $||f(t,x,z) - f(t,y,z)|| \le M||x - y||,\ \forall t,x,y,z \in [0,T] \times H \times H \times H.$

Then for $u_0 \in D(A(0))$ and $y_0 \in H$, there are a BVC mapping $u : [0,T] \to H$ and a $W_{BV}^{1,1}([0,T],H)$ mapping $y : [0,T] \to H$ satisfying

$$y(t) = y_0 + \int_0^t u(s)ds,\ \ t \in [0,T],$$

$$-\frac{du}{dr}(t) \in A(t)u(t) + f(t, u(t), y(t))\ dr\text{-a.e. } t \in [0, T],$$

$$u(0) = u_0$$

with the property: $|u(t) - u(\tau)| \leq K|r(t) - r(\tau)|$ *for all* $t, \tau \in [0, T]$ *for some constant* $K \in]0, \infty[$.

Proof By [8, Theorem 3.1] and the assumptions on f, for any continuous mapping $h : [0, T] \to H$, there is a unique BVC solution v_h to the inclusion

$$\begin{cases} v_h(0) = u_0 \in D(A(0)) \\ -\dfrac{dv_h}{dr}(t) \in A(t)v_h(t) + f(t, v_h(t), h(t))\ dr\text{-a.e.} \end{cases}$$

with $||v_h(t)|| \leq K,\ t \in [0, T]$ and $||v_h(t) - v_h(\tau)|| \leq K(r(t) - r(\tau)),\ t, \tau \in [0, T]$ so that

$$dv_h = \frac{dv_h}{dr} dr$$

with $\frac{dv_h}{dr} \in K\overline{B}_H$, consequently $\frac{dv_h}{dr} \in L^\infty_H([0, T], dr)$. Let consider the closed convex subset $\mathscr{X}$ in the Banach space $\mathscr{C}_H([0, T])$ defined by

$$\mathscr{X} := \{u : [0, T] \to H : u(t) = u_0 + \int_0^t \dot{u}(s)ds,\ \dot{u} \in S^1_{K\overline{B}_H},\ t \in [0, T]\}$$

where $S^1_{K\overline{B}_H}$ denotes the set of all integrable selections of the convex weakly compact valued constant multifunction $K\overline{B}_H$. Now for each $h \in \mathscr{X}$, let us consider the mapping

$$\Phi(h)(t) := u_0 + \int_0^t v_h(s)ds,\ t \in [0, T].$$

Then it is clear that $\Phi(h) \in \mathscr{X}$. Our aim is to prove the existence theorem by applying some ideas developed in Castaing et al. [24] via a generalized fixed point theorem [36, 44]. Nevertheless this needs a careful look using the estimation of the BVC solution given above. For this purpose, we first claim that $\Phi : \mathscr{X} \to \mathscr{X}$ is continuous and for any $h \in \mathscr{X}$ and for any $t \in [0, T]$ the inclusion holds

$$\Phi(h)(t) \in u_0 + \int_0^t \overline{co}[D(A(s)) \cap K\overline{B}_H]ds.$$

Since $s \mapsto \overline{co}[D(A(s)) \cap K\overline{B}_H]$ is a convex compact valued and integrably bounded multifunction using the ball-compactness assumption, the second member is convex compact valued [14] so that $\Phi(\mathscr{X})$ is equicontinuous and relatively compact in the Banach space $\mathscr{C}_H([0,T])$. Now we check that Φ is continuous. It is sufficient to show that, if (h_n) converges uniformly to h in $\mathscr{X}$, then BVC solution v_{h_n} associated with h_n

$$\begin{cases} v_{h_n}(0) = u_0 \in D(A(0)) \\ -\dfrac{dv_{h_n}}{dr}(t) \in A(t)v_{h_n}(t) + f(t, v_{h_n}(t), h_n(t)) \ dr\text{-a.e.} \end{cases}$$

pointwisely converges to the BVC solution v_h associated with h

$$\begin{cases} v_h(0) = u_0 \in D(A(0)) \\ -\dfrac{dv_h}{dr}(t) \in A(t)v_h(t) + f(t, v_h(t), h(t)) \ dr\text{-a.e.} \end{cases}$$

As $D(A(t))$ is ball compact, (v_{h_n}) is uniformly bounded, and bounded in variation since $||v_{h_n}(t) - v_{h_n}(\tau)|| \leq K(r(t) - r(\tau)),\ t, \tau \in [0,T]$, we may assume that (v_{h_n}) pointwisely converges to a BVC mapping v. As $v_{h_n} = v_0 + \int_{]0,t]} \frac{dv_{h_n}}{dr} dr,\ t \in [0,T]$ and $\frac{dv_{h_n}}{dr}(s) \in K\overline{B}_H,\ s \in [0,T]$, we may assume that $(\frac{dv_{h_n}}{dr})$ converges weakly in $L^1_H([0,T], dr)$ to $w \in L^1_H([0,T], dr)$ with $w(t) \in K\overline{B}_H,\ t \in [0,T]$ so that

$$\text{weak-}\lim_n v_{h_n} = u_0 + \int_{]0,t]} w dr := z(t),\ t \in [0,T].$$

By identifying the limits, we get

$$v(t) = z(t) = u_0 + \int_{]0,t]} w dr$$

with $\frac{dv}{dr} = w$ so that $\lim_n f(t, v_{h_n}(t), h_n(t)) = f(t, v(t), h(t)),\ t \in [0,T]$. Consequently we may assume that $(\frac{dv_{h_n}}{dr} + f(., v_{h_n}(.), h_n(.)))$ Komlos converges to $\frac{dv}{dr} - f(., v(.), h(.))$. For simplicity, set $g_n(t) = f(t, v_{h_n}(t), h_n(t))$ and $g(t) = f(t, v(t), h(t))$. There is a dr-negligible set N such that for $t \in I \setminus N$ and

$$\lim_{n\to\infty} \frac{1}{n} \sum_{j=1}^{n} \left(\frac{dv_{h_j}}{dr}(t) + g_j(t) \right) = \frac{dv}{dr}(t) + g(t).$$

Let $\eta \in D(A(t))$. From

$$\left\langle \frac{dv_{h_n}}{dr}(t) + g_n(t), v(t) - \eta \right\rangle$$
$$= \left\langle \frac{dv_{h_n}}{dr}(t) + g_n(t), v_{h_n}(t) - \eta \right\rangle + \left\langle \frac{dv_{h_n}}{dr}(t) + g_n(t), v(t) - v_{h_n}(t) \right\rangle,$$

let us write

$$\frac{1}{n}\sum_{j=1}^{n}\left\langle\frac{dv_{h_j}}{dr}(t)+g_j(t), v(t)-\eta\right\rangle$$
$$=\frac{1}{n}\sum_{j=1}^{n}\left\langle\frac{dv_{h_j}}{dr}(t)+g_j(t), v_{h_j}(t)-\eta\right\rangle+\frac{1}{n}\sum_{j=1}^{n}\left\langle\frac{dv_{h_j}}{dr}(t)+g_j(t), v(t)-v_{h_j}(t)\right\rangle,$$

so that

$$\frac{1}{n}\sum_{j=1}^{n}\left\langle\frac{dv_{h_j}}{dr}(t)+g_j(t), v(t)-\eta\right\rangle$$
$$\leq\frac{1}{n}\sum_{j=1}^{n}\left\langle A^0(t,\eta), \eta-v_{h_j}(t)\right\rangle+\text{(Constant)}\frac{1}{n}\sum_{j=1}^{n}\|v(t)-v_{h_j}(t))\|.$$

Passing to the limit when $n\to\infty$, this last inequality gives immediately

$$\left\langle\frac{dv}{dr}(t)+g(t), v(t)-\eta\right\rangle\leq\left\langle A^0(t,\eta), \eta-v(t)\right\rangle \text{ a.e.}$$

As a consequence, by Lemma 2.2, $-\frac{dv}{dr}(t)\in A(t)v(t)+g(t)=A(t)v(t)+f(t,v(t),h(t))$ a.e. with $v(0)=u_0\in D(A(0))$ so that by uniqueness $v=v_h$. Now let us check that $\Phi:\mathscr{X}\to\mathscr{X}$ is continuous. Let $h_n\to h$. We have

$$\Phi(h_n)(t)-\Phi(h)(t)=\int_0^t v_{h_n}(s)ds-\int_0^t v_h(s)ds=\int_0^t[v_{h_n}(s)-v_h(s)]ds$$

As $||v_{h_n}(.)-v_h(.)||\to 0$ pointwisely and is uniformly bounded : $||v_{h_n}(.)-v_h(.)||\leq 2K$, by we conclude that

$$\sup_{t\in[0,T]}||\Phi(h_n)(t)-\Phi(h)(t)||\leq\sup_{t\in[0,T]}\int_0^t||v_{h_n}(.)-v_h(.)||ds\to 0$$

so that $\Phi(h_n)-\Phi(h)\to 0$ in $\mathscr{C}_H([0,T])$. Here one may invoke a general fact that on bounded subsets of L^∞, the topology of convergence in measure coincides with the topology of uniform convergence on uniformly integrable sets, i.e., on relatively weakly compact subsets, alias the Mackey topology. This is a lemma due to Grothendieck [33, Ch.5 §4 no 1 Prop. 1 and exercice] (see also [15] for a more general result concerning the Mackey topology for bounded sequences in $L^\infty_{E^*}$). Since $\Phi:\mathscr{X}\to\mathscr{X}$ is continuous and $\Phi(\mathscr{X})$ is relatively compact in $\mathscr{C}_H([0,T])$, by [36, 44] Φ has a fixed point, say $h=\Phi(h)\in\mathscr{X}$, that means

$$h(t) = \Phi(h)(t) = u_0 + \int_0^t v_h(s)ds, \ t \in [0, T],$$

$$\begin{cases} v_h(0) = u_0 \in D(A(0)) \\ -\dfrac{dv_h}{dr}(t) \in A(t)v_h(t) + f(t, v_h(t), h(t)) \ dr\text{-a.e.} \end{cases}$$

The proof is complete.

The following results are sharp variants of Theorem 3.3.

Theorem 3.4 *Let for every $t \in [0, T]$, $A(t) : D(A(t)) \subset H \to 2^H$ be a maximal monotone operator with $D(A(t)$ ball compact for every $t \in [0, T]$ satisfying $(H2)$ and*

$(H1)'$ *there exists a function $\beta \in W^{1,1}([0, T], \mathbb{R}; dt)$ which is nonnegative on $[0, T]$ and non-decreasing with $\beta(T) < \infty$ such that*

$$\text{dis}(A(t), A(s)) \leq |\beta(t) - \beta(s)|, \ \forall s, t \in [0, T].$$

$(H1)^*$ *For any $t \in [0, T]$ and for any $x \in D(A(t))$, $A(t)x$ is cone-valued.*

Let $f : [0, T] \times H \times H \to H$ be such that for every $x, y \in H \times H$ the mapping $f(., x, y)$ is Lebesgue-measurable on $[0, T]$ and for every $t \in [0, T]$, $f(t, ., .)$ is continuous on $H \times H$ and satisfying

(i) $||f(t, x, y)|| \leq M(1 + ||x||), \ \forall t, x, y \in [0, T] \times H \times H.$
(ii) $||f(t, x, z) - f(t, y, z)|| \leq M||x - y||, \ \forall t, x, y, z \in [0, T] \times H \times H \times H.$

Then, for all $u_0 \in D(A(0))$, $y_0 \in H$, there are an absolutely continuous mapping $u : [0, T] \to H$ and an absolutely continuous mapping $y : [0, T] \to H$ satisfying

$$y(t) = y_0 + \int_0^t u(s)ds, \quad t \in [0, T],$$

$$-\frac{du}{dt}(t) \in A(t)u(t) + f(t, u(t), y(t)) \ dt - \text{a.e. } t \in [0, T], u(0) = u_0,$$

with

$$\|\dot{u}(t)\| \leq \big(K + M(1 + K)\big)(\dot{\beta}(t) + 1) + M(1 + K)$$

for a.e. $t \in [0, T]$, for some positive constant K.

Proof By [7, Theorem 3.4] and the assumptions on f, for any continuous mapping $h : [0, T] \to H$, there is a unique AC solution v_h to the inclusion

$$\begin{cases} v_h(0) = u_0 \in D(A(0)) \\ -\dot{v}_h(t) \in A(t)v_h(t) + f(t, v_h(t), h(t)) \ dt\text{-a.e.} \end{cases}$$

with $||\dot{v}_h(t)|| \leq \gamma(t) := (K + M(1+K))(\dot{\beta}(t)+1) + M(1+K)$ a.e. $t \in [0, T]$ so that $\gamma \in L^1_{\mathbb{R}}([0, T])$ and $||v_h(t)|| \leq L = \text{Constant}, \ t \in [0, T]$. Let us consider the closed convex subset $\mathscr{X}$ in the Banach space $\mathscr{C}_H([0, T])$ defined by

$$\mathscr{X} := \{u : [0, T] \to H : u(t) = u_0 + \int_0^t \dot{u}(s)ds, \ \dot{u} \in S^1_{L\overline{B}_H}, \ t \in [0, T]\}$$

where $S^1_{L\overline{B}_H}$ denotes the set of all integrable selections of the convex weakly compact valued constant multifunction $L\overline{B}_H$. Now for each $h \in \mathscr{X}$, let us consider the mapping

$$\Phi(h)(t) := u_0 + \int_0^t v_h(s)ds, \ t \in [0, T].$$

Then it is clear that $\Phi(h) \in \mathscr{X}$. Our aim is to prove the existence theorem by applying some ideas developed in Castaing et al. [24] via a generalized fixed point theorem [36, 44]. Nevertheless this needs a careful look using the estimation of the AC solution given above. For this purpose, we first claim that $\Phi : \mathscr{X} \to \mathscr{X}$ is continuous for any $h \in \mathscr{X}$ and for any $t \in [0, T]$, the inclusion holds

$$\Phi(h)(t) \in u_0 + \int_0^t \overline{co}[D(A(s)) \cap L\overline{B}_H]ds.$$

Since $s \mapsto \overline{co}[D(A(s)) \cap L\overline{B}_H]$ is a convex compact valued and integrably bounded multifunction, the second member is convex compact valued [14] so that $\Phi(\mathscr{X})$ is equicontinuous and relatively compact in the Banach space $\mathscr{C}_H([0, T])$. Now we check that Φ is continuous. It is sufficient to show that, if h_n converges uniformly to h in $\mathscr{X}$, then the AC solution v_{h_n} associated with h_n

$$\begin{cases} v_{h_n}(0) = u_0 \in D(A(0)) \\ -\dot{v}_{h_n}(t) \in A(t)v_{h_n}(t) + f(t, v_{h_n}(t), h_n(t)) \ dt\text{-a.e.} \end{cases}$$

converges uniformly to the AC solution v_h associated with h

$$\begin{cases} v_h(0) = u_0 \in D(A(0)) \\ -\dot{v}_h(t) \in A(t)v_h(t) + f(t, v_h(t), h(t)) \ dt\text{-a.e.} \end{cases}$$

We have

$$-\dot{v}_{h_n}(t) \in A(t)v_{h_n}(t) + f(t, v_{h_n}(t), h_n(t)), \ \text{a.e. } t \in [0, T],$$

with the estimation $||\dot{v}_{h_n}(t)|| \leq \gamma(t)$ and $\gamma \in L^1_{\mathbb{R}}([0, T])$ for all $n \in \mathbb{N}$. As $D(A(t))$ is ball compact and $(\dot{v}_{h_n})$ is relatively weakly compact in

$L^1_H([0,T])$, we may assume that (v_{h_n}) converges uniformly to an absolutely continuous mapping v such that $v(t) = u_0 + \int_0^t \dot{v}(s)ds$, $t \in [0,T]$, $||\dot{v}(t)|| \leq \gamma(t)$, $t \in [0,T]$, and $(\dot{v}_{h_n})\,\sigma(L^1_H, L^\infty_H)$ converges to $\dot{v}$ so that $\lim_n f(t, v_{h_n}(t), h_n(t)) = f(t, v(t), h(t))$, $t \in [0,T]$. Consequently we may assume that $(\dot{v}_{h_n} + f(., v_{h_n}(.), h_n(.)))$ Komlos converges to $\dot{v} - f(., v(.), h(.))$. Let us set $g_n(t) = f(t, v_{h_n}(t), h_n(t))$ and $g(t) = f(t, v(t), h(t))$. There is a negligible set N such that for $t \in [0,T] \setminus N$ and

$$\lim_{n\to\infty} \frac{1}{n}\sum_{j=1}^{n} \big(\dot{v}_{h_j}(t) + g_j(t)\big) = \dot{v}(t) + g(t).$$

Let $\eta \in D(A(t))$. From

$$\begin{aligned}&\big\langle \dot{v}_{h_n}(t) + g_n(t), v(t) - \eta\big\rangle \\ &\qquad = \big\langle \dot{v}_{h_n}(t) + g_n(t), v_{h_n}(t) - \eta\big\rangle + \big\langle \dot{v}_{h_n}(t) + g_n(t), v(t) - v_{h_n}(t)\big\rangle\end{aligned}$$

let us write

$$\begin{aligned}&\frac{1}{n}\sum_{j=1}^{n}\big\langle \dot{v}_{h_j}(t) + g_j(t), v(t) - \eta\big\rangle \\ &= \frac{1}{n}\sum_{j=1}^{n}\big\langle \dot{v}_{h_j}(t) + g_j(t), v_{h_j}(t) - \eta\big\rangle + \frac{1}{n}\sum_{j=1}^{n}\big\langle \dot{v}_{h_j}(t) + g_j(t), v(t) - v_{h_j}(t)\big\rangle,\end{aligned}$$

so that

$$\begin{aligned}&\frac{1}{n}\sum_{j=1}^{n}\big\langle \dot{v}_{h_j}(t) + g_j(t), v(t) - \eta\big\rangle \\ &\quad\leq \frac{1}{n}\sum_{j=1}^{n}\big\langle A^0(t,\eta), \eta - v_{h_j}(t)\big\rangle + \big(\gamma(t) + \text{Constant}\big)\frac{1}{n}\sum_{j=1}^{n}\|v(t) - v_{h_j}(t))\|.\end{aligned}$$

Passing to the limit when $n \to \infty$, this last inequality gives immediately

$$\big\langle \dot{v}(t) + g(t), v(t) - \eta\big\rangle \leq \big\langle A^0(t,\eta), \eta - v(t)\big\rangle \text{ a.e.}$$

As a consequence, $-\dot{v}(t) \in A(t)v(t) + g(t) = A(t)v(t) + f(t, v(t), h(t))$ a.e. with $v(0) = u_0 \in D(A(0))$ so that by uniqueness $v = v_h$. Since $\Phi : \mathscr{X} \to \mathscr{X}$ is continuous and $\Phi(\mathscr{X})$ is relatively compact in $\mathscr{C}_H([0,T])$, by [36, 44] Φ has a fixed point, say $h = \Phi(h) \in \mathscr{X}$, that means

$$h(t) = \Phi(h)(t) = u_0 + \int_0^t v_h(s)ds, \ t \in [0, T],$$

$$\begin{cases} v_h(0) = u_0 \in D(A(0)) \\ -\dot{v}_h(t) \in A(t)v_h(t) + f(t, v_h(t), h(t)) \ dt\text{-a.e.} \end{cases}$$

The proof is complete.

Comments The use of a generalized fixed point theorem is initiated in [24] dealing with some second-order sweeping process associated with a closed moving set $C(t, u)$. Actually it is possible to obtain a variant of Theorem 3.4 by assuming that $A(t) : D(A(t)) \subset H \to 2^H$ is a maximal monotone operator with $D(A(t)$ ball compact for every $t \in [0, T]$ satisfying $(H2)$ and
$(H1)'$ there exists a function $\beta \in W^{1,2}([0, T], \mathbb{R}; dt)$ which is nonnegative on I and non-decreasing with $\beta(T) < \infty$ such that

$$\text{dis}(A(t), A(s)) \le |\beta(t) - \beta(s)|, \ \forall s, t \in [0, T].$$

Here using fixed point theorem provides a short proof with new approach involving the continuous dependance of the trajectory v_h associated with the control $h \in \mathscr{X}$ and also the compactness of the integral of convex compact integrably bounded multifunctions [14].

4 Evolution Problems with Lipschitz Variation Maximal Monotone Operator and Application to Viscosity and Control

Now, based on the existence and uniqueness of $W_H^{2,\infty}([0, T])$ solution to evolution inclusion

$$(\mathscr{S}_1) \begin{cases} 0 \in \ddot{u}(t) + A(t)\dot{u}(t) + f(t, u(t)), \ t \in [0, T] \\ u(0) = u_0, \dot{u}(0) = \dot{u}_0 \in D(A(0)) \end{cases}$$

we will present some problems in optimal control in a second-order evolution inclusion driven by a Lipschitz variation maximal monotone operator $A(t)$ in the same vein as in Castaing-Marques-Raynaud de Fitte [25] dealing with the sweeping process. Before going further, we note that $(\mathscr{S}_1)$ contains the evolution problem associated with the sweeping process by a closed convex Lipschitzian mapping $C : [0, T] \to \text{cc}(H)$

$$\begin{cases} 0 \in \ddot{u}(t) + N_{C(t)}(\dot{u}(t)) + f(t, u(t)), \ t \in [0, T] \\ u(0) = u_0, \dot{u}(0) = \dot{u}_0 \in C(0) \end{cases}$$

by taking $A(t) = \partial\Psi_{C(t)}$ in $(\mathscr{S}_1)$.

We need some notations and background on Young measures in this special context. For the sake of completeness, we summarize some useful facts concerning Young measures. Let $(\Omega, \mathcal{F}, P)$ be a complete probability space. Let X be a Polish space, and let $\mathcal{C}^b(X)$ be the space of all bounded continuous functions defined on X. Let $\mathcal{M}_+^1(X)$ be the set of all Borel probability measures on X equipped with the narrow topology. A Young measure $\lambda : \Omega \to \mathcal{M}_+^1(X)$ is, by definition, a *scalarly measurable* mapping from Ω into $\mathcal{M}_+^1(X)$, that is, for every $f \in \mathcal{C}^b(X)$, the mapping $\omega \mapsto \langle f, \lambda_\omega \rangle := \int_X f(x)\, d\lambda_\omega(x)$ is $\mathcal{F}$-measurable. A sequence (λ^n) in the space of Young measures $\mathcal{Y}(\Omega, \mathcal{F}, P; \mathcal{M}_+^1(X))$ *stably converges* to a Young measure $\lambda \in \mathcal{Y}(\Omega, \mathcal{F}, P; \mathcal{M}_+^1(X))$ if the following holds:

$$\lim_{n\to\infty} \int_A \left[\int_X f(x)\, d\lambda_\omega^n(x)\right] dP(\omega) = \int_A \left[\int_X f(x)\, d\lambda_\omega(x)\right] dP(\omega)$$

for every $A \in \mathcal{F}$ and for every $f \in \mathcal{C}^b(X)$. We recall and summarize some results for Young measures.

Theorem 4.5 ([22, Theorem 3.3.1]) *Assume that S and T are Polish spaces. Let (μ^n) be a sequence in $\mathcal{Y}(\Omega, \mathcal{F}, P; \mathcal{M}_+^1(S))$, and let (ν^n) be a sequence in $\mathcal{Y}(\Omega, \mathcal{F}, P; \mathcal{M}_+^1(T))$. Assume that*

(i) *(μ^n) converges in probability to $\mu^\infty \in \mathcal{Y}(\Omega, \mathcal{F}, P; \mathcal{M}_+^1(S))$,*
(ii) *(ν^n) stably converges to $\nu^\infty \in \mathcal{Y}(\Omega, \mathcal{F}, P; \mathcal{M}_+^1(T))$.*

Then $(\mu^n \otimes \nu^n)$ stably converges to $\mu^\infty \otimes \nu^\infty$ in $\mathcal{Y}(\Omega, \mathcal{F}, P; \mathcal{M}_+^1(S \times T))$.

Theorem 4.6 ([22, Theorem 6.3.5]) *Assume that X and Z are Polish spaces. Let (u^n) be sequence of $\mathcal{F}$-measurable mappings from Ω into X such that (u^n) converges in probability to a $\mathcal{F}$-measurable mapping u^∞ from Ω into X, and let (v^n) be a sequence of $\mathcal{F}$-measurable mappings from Ω into Z such that (v^n) stably converges to $v^\infty \in \mathcal{Y}(\Omega, \mathcal{F}, P; \mathcal{M}_+^1(Z))$. Let $h : \Omega \times X \times Z \to \mathbb{R}$ be a Carathéodory integrand such that the sequence $(h(., u_n(.), v_n(.))$ is uniformly integrable. Then the following holds:*

$$\lim_{n\to\infty} \int_\Omega h(\omega, u^n(\omega), v^n(\omega))\, dP(\omega) = \int_\Omega \left[\int_Z h(\omega, u^\infty(\omega), z)\, dv_\omega^\infty(z)\right] dP(\omega).$$

In the remainder, Z is a compact metric space, and $\mathcal{M}_+^1(Z)$ is the space of all probability Radon measures on Z. We will endow $\mathcal{M}_+^1(Z)$ with the narrow topology so that $\mathcal{M}_+^1(Z)$ is a compact metrizable space. Let us denote by $\mathcal{Y}([0,T]; \mathcal{M}_+^1(Z))$ the space of all Young measures (alias *relaxed controls*) defined on $[0,T]$ endowed with the stable topology so that $\mathcal{Y}([0,T]; \mathcal{M}_+^1(Z))$ is a compact metrizable space with respect to this topology. By the Portmanteau Theorem for Young measures [22, Theorem 2.1.3], a sequence (ν^n) in $\mathcal{Y}([0,T]; \mathcal{M}_+^1(Z))$ stably converges to $\nu \in \mathcal{Y}([0,T]; \mathcal{M}_+^1(Z))$ if

$$\lim_{n\to\infty} \int_0^T \left[\int_Z h_t(z) d\nu_t^n(z)\right] dt = \int_0^T \left[\int_Z h_t(z) d\nu_t(z)\right] dt$$

for all $h \in L^1_{\mathscr{C}(Z)}([0, T])$, where $\mathscr{C}(Z)$ denotes the space of all continuous real-valued functions defined on Z endowed with the norm of uniform convergence. Finally let us denote by $\mathscr{Z}$ the set of all Lebesgue-measurable mappings (alias *original controls*) $z : [0, T] \to Z$ and $\mathscr{R} := \mathscr{Y}([0, T]; \mathscr{M}^1_+(Z))$ the set of all relaxed controls (alias Young measures) associated with Z. In the remainder, we assume that $H = \mathbb{R}^d$ and Z is a compact subset in H.

For simplicity, let us consider a mapping $f : [0, T] \times H \to H$ satisfying

(i) for every $x \in H \times Z$, $f(., x)$ is Lebesgue-measurable on $[0, T]$,
(ii) there is $M > 0$ such that

$$||f(t, x)|| \leq M(1 + ||x||)$$

for all (t, x) in $[0, T] \times H$, and

$$||f(t, x) - f(t, y)|| \leq M||x - y||$$

for all $(t, x, y) \in [0, T] \times H \times H$.

We consider the $W^{2,\infty}_H([0, T])$ solution set of the two following control problems

$$(\mathscr{S}_{\mathscr{O}}) \begin{cases} 0 \in \ddot{u}_{x,y,\zeta}(t) + A(t)\dot{u}_{x,y,\zeta}(t)) + f(t, u_{x,y,\zeta}(t)) + \zeta(t), \ t \in [0, T] \\ u_{x,y,\zeta}(0) = x \in H, \dot{u}_{x,y,\zeta}(0) = y \in D(A(0)) \end{cases}$$

and

$$(\mathscr{S}_{\mathscr{R}}) \begin{cases} 0 \in \ddot{u}_{x,y,\lambda}(t) + A(t)\dot{u}_{x,y,\lambda}(t)) + f(t, u_{x,y,\lambda}(t)) + \mathrm{bar}(\lambda_t), \ t \in [0, T] \\ u_{x,y,\lambda}(0) = x \in H, \dot{u}_{x,y,\lambda}(0) = y \in D(A(0)) \end{cases}$$

where ζ belongs to the set $\mathscr{Z}$ of all Lebesgue-measurable mappings (alias original controls) $\zeta : [0, T] \to Z$ original and λ belongs to the set $\mathscr{R}$ of all relaxed controls. Taking $(\mathscr{S}_1)$ into account, for each $(x, y, \zeta) \in H \times D(A(0)) \times \mathscr{Z}$ (resp. $(x, y, \lambda) \in H \times D(A(0)) \times \mathscr{R}$, there exists a unique $W^{2,\infty}_H(]0, T])$ solutions, solution $u_{x,y,\zeta}$ (resp. $u_{x,y,\lambda}$), to $(\mathscr{S}_{\mathscr{O}})$ (resp. $(\mathscr{S}_{\mathscr{R}})$). We aim to present some problems in the framework of optimal control theory for the above inclusions. In particular, we state a viscosity property of the value function associated with these evolution inclusions. Similar problems driven by evolution inclusion with perturbation containing Young measures are initiated by [22, 23]. However, the present study deals with a new setting in the sense that it concerns a second-order evolution inclusion involving time-dependent maximal monotone operator.

Now we present a lemma which is useful for our purpose.

Lemma 4.6 *Let for all $t \in [0, T]$, $A(t) : D(A(t)) \subset H \to 2^H$ be a maximal monotone operator satisfying $(H1)$ and $(H2)$. Let $f : [0, T] \times H \to H$ be a mapping satisfying*

(i) *for every* $x \in H \times Z$, $f(., x)$ *is Lebesgue-measurable on* $[0, T]$,
(ii) *there is* $M > 0$ *such that*

$$||f(t, x)|| \leq M(1 + ||x||)$$

for all (t, x) *in* $[0, T] \times H$, *and*

$$||f(t, x) - f(t, y)|| \leq M||x - y||$$

for all $(t, x, y) \in [0, T] \times H \times H$.

Let $h_n, h \in L^{\infty}_H([0, T], dt)$ *with* $||h_n(t)|| \leq 1$ *for all* $t \in [0, T]$, *for all* $n \in \mathbb{N}$ *and* $||h(t)|| \leq 1$ *for all* $t \in [0, T]$. *Let us consider the two following second-order evolution inclusions:*

$$\mathscr{S}(A, f, h_n, x, y) \begin{cases} 0 \in \ddot{u}_{x,y,h_n}(t) + A(t)\dot{u}_{x,y,h_n}(t) + f(t, u_{x,y,h_n}(t)) + h_n(t), \ t \in [0, T] \\ u_{x,y,h_n}(0) = x, \dot{u}_{x,y,h_n}(0) = y \in D(A(0)) \end{cases}$$

$$\mathscr{S}(A, f, h, x, y) \begin{cases} 0 \in \ddot{u}_{x,y,h}(t) + A(t)\dot{u}_{x,y,h}(t) + f(t, u_{x,y,h}(t)) + h(t), \ t \in [0, T] \\ u_{x,y,h}(0) = x, \dot{u}_{x,y,h}(0) = y \in D(A(0)) \end{cases}$$

where u_{x,y,h_n} *(resp.* $u_{x,y,h}$*) is the unique* $W^{2,\infty}_H([0, T])$ *solution to* $(\mathscr{S}(A, f, h_n, x, y))$ *(resp.* $(\mathscr{S}(A, f, h_n, x, y))$*). Assume that* (h_n) $\sigma(L^1, L^{\infty})$ *converges to* h. *Then* (u_{x,y,h_n}) *converges pointwisely to* $u_{x,y,h}$.

Proof We note that $\ddot{u}_{x,y,h_n}$ is *uniformly bounded*, so there is $u \in W^{2,\infty}_H([0, T])$ such that

$u_{x,y,h_n} \to u$ pointwisely with $u(0) = x$,
$\dot{u}_{x,y,h_n} \to \dot{u}$ pointwisely with $\dot{u}(0) = y$,
$\ddot{u}_{x,y,h_n} \to \ddot{u}$ with respect to $\sigma(L^1, L^{\infty})$.

Using Lemma 2.3, it is not difficult to see that $\dot{u}(t) \in D(A(t))$ for every $t \in [0, T]$. As $f(t, u_{x,y,h_n}(t)) \to f(t, u(t))$ pointwisely so that $f(., u_{x,y,h_n}(.)) \to f(.., u(.))$ with respect to $\sigma(L^1, L^{\infty})$. Since (h_n) $\sigma(L^1, L^{\infty})$ converges to h, so that $f(., u_{x,y,h_n}(.)) + h_n \to f(t., u(.)) + h$ with respect to $\sigma(L^1, L^{\infty})$. And so $\ddot{u}_{x,y,h_n}(.) + f(., u_{x,y,h_n}(.)) + h_n(.)$ $\sigma(L^1, L^{\infty})$ converges to $\dot{u} + f(.., u(.)) + h$. As a consequence, we may also assume that $\ddot{u}_{x,y,h_n}(.) + f(., u_{x,y,h_n}(.)) + h_n(.)$ Komlos converges to $\dot{u}+f(.., u(.))+h$. Coming back to the inclusion $-\ddot{u}_{x,y,h_n}(t) - f(t, u_{x,y,h_n}(t)) - h_n(t) \in A(t)\dot{u}_{x,y,h_n}(t)$, we have by the monotonicity of $A(t)$

$$\langle \ddot{u}_{x,y,h_n}(t) + f(t, u_{x,y,h_n}(t)) + h_n(t), \dot{u}_{x,y,h_n}(t) - \eta \rangle \leq \langle A^0(t, \eta), \eta - \dot{u}_{x,y,h_n}(t) \rangle$$

for any $\eta \in D(A(t))$. For notational convenience, set

$$v_n(t) = \ddot{u}_{x,y,h_n}(t) + f(t, u_{x,y,h_n}(t)) + h_n(t), \forall t \in [0, T],$$
$$v(t) = \ddot{u}(t) + f(t, u(t)) + h(t), \forall t \in [0, T].$$

There is a negligible set N such that

$$\lim_n \frac{1}{n}\sum_{i=1}^{n} v_i(t) = v(t)$$

for $t \notin N$. Let us write

$$\langle v_n(t), \dot{u}(t) - \eta\rangle = \langle v_n(t), \dot{u}_{x,y,h_n}(t) - \eta\rangle + \langle v_n(t), \dot{u}(t) - \dot{u}_{x,y,h_n}(t)\rangle$$

so that

$$\frac{1}{n}\sum_{i=1}^{n}\langle v_i(t), \dot{u}(t)-\eta\rangle=\frac{1}{n}\sum_{i=1}^{n}\langle v_i(t), \dot{u}_{x,y,h_i}(t)-\eta\rangle+\frac{1}{n}\sum_{i=1}^{n}\langle v_i(t), \dot{u}(t)-\dot{u}_{x,y,h_i}(t)\rangle$$

$$\leq\frac{1}{n}\sum_{i=1}^{n}\langle A^0(t,\eta), \eta-\dot{u}_{x,y,h_i}(t)\rangle+L\frac{1}{n}\sum_{i=1}^{n}||\dot{u}(t)-\dot{u}_{x,y,h_i}(t)||,$$

where L is a positive generic constant. Passing to the limit when n goes to ∞ in this inequality gives immediately

$$\langle v(t), \dot{u}(t) - \eta\rangle \leq \langle A^0(t,\eta), \eta - \dot{u}(t)\rangle$$

so that by Lemma 2.2 we get

$$-\ddot{u}(t) - f(t, u_{x,y,h}(t)) - h(t) \in A(t)\dot{u}(t) \text{ a.e.}$$

with $u(0) = x$ and $\dot{u}(0) = y$. Due to the uniqueness of solution, we get $u(t) = u_{x,y,h}(t)$ for all $t \in [0, T]$. The proof is complete.

The following shows the continuous dependence of the solution with respect to the control.

Theorem 4.7 *Let for all* $t \in [0, T]$, $A(t) : D(A(t)) \subset H \to 2^H$ *be a maximal monotone operator satisfying* $(H1)$ *and* $(H2)$. *Let* $f : [0, T] \times H \to H$ *be a mapping satisfying*

(i) *for every* $x \in H \times Z$, $f(., x)$ *is Lebesgue-measurable on* $[0, T]$,
(ii) *there is* $M > 0$ *such that*

$$||f(t, x)|| \leq M(1 + ||x||)$$

for all (t, x) *in* $[0, T] \times H$, *and*

$$||f(t, x_1) - f(t, x_2)|| \leq M||x_1 - x_2||$$

for all $(t, x_1, (t, x_2,) \in [0, T] \times H \times H$.

Let Z be a compact subset of H. Let us consider the control problem

$$\begin{cases} 0 \in \ddot{u}_{x,y,\nu}(t) + A(t)\dot{u}_{x,y,\nu}(t) + f(t, u_{x,y,\nu}(t)) + \mathrm{bar}(\nu_t), \ t \in [0, T] \\ u_{x,y,\nu}(0) = x, \dot{u}_{x,y,\nu}(0) = y \in D(A(0)) \end{cases}$$

where $\mathrm{bar}(\nu_t)$ *denotes the barycenter of the measure* $\nu_t \in \mathcal{M}_+^1(Z)$ *and* $u_{x,y,\nu}$ *is the unique* $W_H^{2,\infty}([0, T])$ *solution associated with to* $\mathrm{bar}(\nu_t)$. *Then, for each* $t \in [0, T]$, *the mapping* $\nu \mapsto u_{x,y,\nu}$ *is continuous from* $\mathcal{R}$ *to* $C_H([0, T]$, *where* $\mathcal{R}$ *is endowed with the stable topology and* $C_H([0, T]$ *is endowed with the topology of pointwise convergence.*

Proof (a) Let $\nu \in \mathcal{R}$ and let $\mathrm{bar}(\nu) : t \mapsto \mathrm{bar}(\nu_t)$, $t \in [0, T]$. It is easy to check that $\nu \mapsto \mathrm{bar}(\nu)$ from $\mathcal{R}$ to $L_H^\infty([0, T])$ is continuous with respect to the stable topology and the $\sigma(L_H^1, L_H^\infty)$, respectively. Note that $\mathcal{R}$ is compact metrizable for the stable topology. Now let (ν^n) be a sequence in $\mathcal{R}$ which stably converges to $\nu \in \mathcal{R}$. Then $\mathrm{bar}(\nu^n)$ $\sigma(L_H^1, L_H^\infty)$ converges to $\mathrm{bar}(\nu)$. By Lemma 4.6, we see that u_{x,y,ν^n} pointwisely converges to $u_{x,y,\nu}$. The proof is complete.

We are now able to relate the Bolza type problems associated with the maximal monotone operator $A(t)$ as follows:

Theorem 4.8 *With the hypotheses and notations of Theorem* 4.7, *assume that* $J : [0, T] \times H \times Z \to \mathbb{R}$ *is a Carathéodory integrand, that is,* $J(t, ., .)$ *is continuous on* $H \times Z$ *for every* $t \in [0, T]$ *and* $J(., x, z)$ *is Lebesgue-measurable on* $[0, T]$ *for every* $(x, z) \in H \times Z$, *which satisfies the condition* ($\mathcal{C}$)*: for every sequence* (ζ_n) *in* $\mathcal{Z}$, *the sequence* $(J(., u_{x,y,\zeta^n}(.), \zeta^n(.))$ *is uniformly integrable in* $L_{\mathbb{R}}^1([0, T], dt)$, *where* u_{x,y,ζ^n} *denotes the unique* $W_H^{2,\infty}([0, T])$ *solution associated with* ζ^n *to the evolution inclusion*

$$\begin{cases} 0 \in \ddot{u}_{x,y,\zeta^n}(t) + A(t)\dot{u}_{x,y,\zeta^n}(t) + f(t, u_{x,y,\zeta^n}(t)) + \zeta^n(t), \ t \in [0, T] \\ u_{x,y,\zeta^n}(0) = x, \dot{u}_{x,y,\zeta^n}(0) = y \in D(A(0)) \end{cases}$$

Let us consider the control problems

$$\inf(P_{\mathcal{Z}}) := \inf_{\zeta \in \mathcal{Z}} \int_0^T J(t, u_{x,y,\zeta}(t), \zeta(t))\, dt$$

and

$$\inf(P_{\mathcal{R}}) := \inf_{\lambda \in \mathcal{R}} \int_0^T \left[\int_Z J(t, u_{x,y,\lambda}(t), z)\, \lambda_t(dz) \right] dt$$

where $u_{x,y,\zeta}$ *(resp.* $u_{x,y,\lambda}$*) is the unique* $W_H^{2,\infty}([0, T])$ *solution associated with* ζ *(resp.* λ*) to*

$$\begin{cases} 0 \in \ddot{u}_{x,y,\zeta}(t) + A(t)\dot{u}_{x,y,\zeta}(t) + f(t, u_{x,y,\zeta}(t)) + \zeta(t), \ t \in [0, T] \\ u_{x,y,\zeta}(0) = x, \dot{u}_{x,y,\zeta}(0) = y \in D(A(0)) \end{cases}$$

and

$$\begin{cases} 0 \in \ddot{u}_{x,y,\lambda}(t) + A(t)\dot{u}_{x,y,\nu}(t) + f(t, u_{x,y,\nu}(t)) + \text{bar}(\lambda_t), \; t \in [0, T] \\ u_{x,y,\lambda}(0) = x, \, \dot{u}_{x,y,\lambda}(0) = y \in D(A(0)) \end{cases}$$

respectively. Then one has

$$\inf(P_{\mathscr{Z}}) = \inf(P_{\mathscr{R}}).$$

Proof Take a control $\lambda \in \mathscr{R}$. By virtue of the denseness with respect to the stable topology of $\mathscr{Z}$ in $\mathscr{R}$, there is a sequence $(\zeta^n)_{n\in\mathbb{N}}$ in $\mathscr{Z}$ such that the sequence $(\delta_{\zeta^n})_{n\in\mathbb{N}}$ of Young measures associated with $(\zeta^n)_{n\in\mathbb{N}}$ stably converges to λ. By Theorem 4.7, the sequence (u_{x,y,ζ^n}) of $W_H^{2,\infty}([0, T])$ solutions associated with ζ^n pointwisely converges to the unique $W_H^{2,\infty}([0, T])$ solution $u_{x,y,\lambda}$. As $(J(t, u_{x,y,\zeta^n}(t), \zeta^n(t)))$ is uniformly integrable by assumption ($\mathscr{C}$), using Theorem 4.6 (or [22, Theorem 6.3.5]), we get

$$\lim_{n\to\infty} \int_0^T J(t, u_{x,y,\zeta^n}(t), \zeta^n(t))\, dt = \int_0^T \left[\int_Z J(t, u_{x,y,\lambda}, z) d\lambda_t(z)\right] dt.$$

As

$$\int_0^T J(t, u_{x,y,\zeta^n}(t), \zeta^n(t))\, dt \geq \inf(P_{\mathscr{Z}})$$

for all $n \in \mathbb{N}$, so is

$$\int_0^T \left[\int_Z J(t, u_{x,y,\lambda}, z) d\lambda_t(z)\right] dt \geq \inf(P_{\mathscr{Z}});$$

by taking the infimum on $\mathscr{R}$ in this inequality, we get

$$\inf(P_{\mathscr{R}}) \geq \inf(P_{\mathscr{O}})$$

As $\inf(P_{\mathscr{O}}) \geq \inf(P_{\mathscr{R}})$, the proof is complete.

In the framework of optimal control, the above considerations lead to the study of the value function associated with the evolution inclusion

$$\begin{cases} 0 \in \ddot{u}_{\tau,x,y,\nu}(t) + A(t)\dot{u}_{\tau,x,y,\nu}(t) + f(t, u_{\tau,x,y,\nu}(t)) + \text{bar}(\nu_t), \\ u_{\tau,x,y,\nu}(\tau) = x, \, \dot{u}_{\tau,x,y,\nu}(\tau) = y \in D(A(\tau)). \end{cases}$$

The following shows that the value function satisfies the dynamic programming principle (DPP).

Theorem 4.9 *(of dynamic programming principle). Assume the hypothesis and notations of Theorem 4.7, and let $x \in E$, $\tau < T$ and $\sigma > 0$ such that $\tau + \sigma < T$. Assume that $J : [0, T] \times H \times Z \to \mathbb{R}$ is bounded and continuous. Let us consider the value function*

$$V_J(\tau, x, y) = \sup_{\nu\in\mathscr{R}} \int_\tau^T \left[\int_Z J(t, u_{\tau,x,y,\nu}(t), z)\nu_t(dz)\right] dt,$$
$$(\tau, x, y) \in [0, T] \times H \times D(A(\tau))$$

where $u_{\tau,x,y,\nu}$ is the $W_H^{2,\infty}([0, T])$ solution to the evolution inclusion defined on $[\tau, T]$ associated with the control $\nu \in \mathscr{R}$ starting from x, y at time τ

$$\begin{cases} 0 \in \ddot{u}_{\tau,x,y,\nu}(t) + A(t)\dot{u}_{\tau,x,y,\nu}(t) + f(t, u_{\tau,x,y,\nu}(t)) + \mathrm{bar}(\nu_t), \\ u_{\tau,x,y,\nu}(\tau) = x, \dot{u}_{\tau,x,y,\nu}(\tau) = y \in D(A(\tau)) \end{cases}$$

Then the following holds:

$$V_J(\tau, x, y) = \sup_{\nu\in\mathscr{R}} \left\{ \int_\tau^{\tau+\sigma} \left[\int_Z J(t, u_{\tau,x,y,\nu}(t), z)\nu_t(dz)\right] dt + V_J(\tau + \sigma, u_{\tau,x,y,\nu}(\tau + \sigma), \dot{u}_{\tau,x,y,\nu}(\tau + \sigma)\right\}$$

with

$$V_J(\tau + \sigma, u_{\tau,x,\nu}(\tau + \sigma), \dot{u}_{\tau,x,\nu}(\tau + \sigma)) = \sup_{\mu\in\mathscr{R}} \int_{\tau+\sigma}^T \left[\int_Z J(t, v_{\tau+\sigma,u_{\tau,x,y,\nu}(\tau+\sigma),\dot{u}_{\tau,x,y,\nu}(\tau+\sigma),\mu}(t), z)\mu_t(dz)\right] dt$$

where $v_{\tau+\sigma,u_{\tau,x,y,\nu}(\tau+\sigma),\dot{u}_{\tau,x,y,\nu}(\tau+\sigma),\mu}$[1] *is the $W_H^{2,\infty}(\tau + \sigma, T)$ solution defined on $[\tau+\sigma, T]$ associated with the control $\mu \in \mathscr{R}$ starting from $u_{\tau,x,\nu}(\tau+\sigma)$, $\dot{u}_{\tau,x,\nu}(\tau+\sigma)$ at time $\tau + \sigma$*

$$\begin{cases} 0 \in v_{\tau+\sigma,u_{\tau,x,y,\nu}(\tau+\sigma),\dot{u}_{\tau,x,y,\nu}(\tau+\sigma),\mu}(t) + A(t)v_{\tau+\sigma,u_{\tau,x,y\nu}(\tau+\sigma),\dot{u}_{\tau,x,y,\nu}(\tau+\sigma),\mu}(t), \\ \quad + f(t, v_{\tau+\sigma,u_{\tau,x,y,\nu}(\tau+\sigma),\dot{u}_{\tau,x,y,\nu}(\tau+\sigma),\mu}(t)) + \mathrm{bar}(\mu_t), \\ v_{\tau+\sigma,u_{\tau,x,y,\nu}(\tau+\sigma),\dot{u}_{\tau,x,y,\nu}(\tau+\sigma),\mu}(\tau + \sigma) = u_{\tau,x,y,\nu}(\tau + \sigma), \\ \dot{v}_{\tau+\sigma,u_{\tau,x,y,\nu}(\tau+\sigma),\dot{u}_{\tau,x,y,\nu}(\tau+\sigma),\mu}(\tau + \sigma) = \dot{u}_{\tau,x,y,\nu}(\tau + \sigma) \in D(A(\tau + \sigma)). \end{cases} \tag{13}$$

[1]It is necessary to write completely the expression of the trajectory $v_{\tau+\sigma,u_{\tau,x,y,\nu}(\tau+\sigma),\dot{u}_{\tau,x,y,\nu}(\tau+\sigma),\mu}$ that depends on $(\nu, \mu) \in \mathscr{R} \times \mathscr{R}$ in order to get the continuous dependence with respect to $\nu \in \mathscr{R}$ of $V_J(\tau + \sigma, u_{\tau,x,y,\nu}(\tau + \sigma))$.

Proof Let

$$W_J(\tau, x, y) := \sup_{\nu\in\mathscr{R}} \left\{ \int_\tau^{\tau+\sigma} \left[\int_Z J(t, u_{\tau,x,y,\nu}(t), z)\nu_t(dz) \right] dt \right.$$
$$\left. + V_J(\tau+\sigma, u_{\tau,x,y,\nu}(\tau+\sigma)) \right\}.$$

For any $\nu \in \mathscr{R}$, we have

$$\int_\tau^T \left[\int_Z J(t, u_{\tau,x,y,\nu}(t), z)\nu_t(dz) \right] dt = \int_\tau^{\tau+\sigma} \left[\int_Z J(t, u_{\tau,x,y,\nu}(t), z)\nu_t(dz) \right] dt$$
$$+ \int_{\tau+\sigma}^T \left[\int_Z J(t, u_{\tau,x,y,\nu}(t), z)\nu_t(dz) \right] dt.$$

By the definition of $V_J(\tau+\sigma, u_{\tau,x,y,\nu}(\tau+\sigma), \dot{u}_{\tau,x,y,\nu}(\tau+\sigma))$, we have

$$V_J(\tau+\sigma, u_{\tau,x,y,\nu}(\tau+\sigma), \dot{u}_{\tau,x,y,\nu}(\tau+\sigma)) \geq \int_{\tau+\sigma}^T \left[\int_Z J(t, u_{\tau,x,y,\nu}(t), z)\nu_t(dz) \right] dt.$$

It follows that

$$\int_\tau^T \left[\int_Z J(t, u_{\tau,x,y,\nu}(t), z)\nu_t(dz) \right] dt \leq \int_\tau^{\tau+\sigma} \left[\int_Z J(t, u_{\tau,x,y,\nu}(t), z)\nu_t(dz) \right] dt$$
$$+ V_J(\tau+\sigma, u_{\tau,x,y,\nu}(\tau+\sigma), \dot{u}_{\tau,x,y,\nu}(\tau+\sigma)).$$

By taking the supremum on $\nu \in \mathscr{R}$ in this inequality, we get

$$V_J(\tau, x, y) \leq \sup_{\nu\in\mathscr{R}} \left\{ \int_\tau^{\tau+\sigma} \left[\int_Z J(t, u_{\tau,x,y,\nu}(t), z)\nu_t(dz) \right] dt \right.$$
$$\left. + V_J(\tau+\sigma, u_{\tau,x,y,\nu}(\tau+\sigma), \dot{u}_{\tau,x,y,\nu}(\tau+\sigma)) \right\}$$
$$= W_J(\tau, x, y).$$

Let us prove the converse inequality.
Main fact: $\nu \mapsto V_J(\tau+\sigma, u_{\tau,x,\nu}(\tau+\sigma), \dot{u}_{\tau,x,\nu}(\tau+\sigma))$ is continuous on $\mathscr{R}$.
Let us focus on the expression of $V_J(\tau+\sigma, u_{\tau,x,\nu}(\tau+\sigma), \dot{u}_{\tau,x,\nu}(\tau+\sigma))$:

$$V_J(\tau+\sigma, u_{\tau,x,\nu}(\tau+\sigma), \dot{u}_{\tau,x,\nu}(\tau+\sigma))$$
$$= \sup_{\mu\in\mathscr{R}} \int_{\tau+\sigma}^T \left[\int_Z J(t, v_{\tau+\sigma, u_{\tau,x,\nu}(\tau+\sigma), \dot{u}_{\tau,x,\nu}(\tau+\sigma),\mu}(t), z)\mu_t(dz) \right] dt$$

where $v_{\tau+\sigma,u_{\tau,x,\nu}(\tau+\sigma),\dot u_{\tau,x,\nu}(\tau+\sigma),\mu}$ denotes the trajectory solution on $[\tau+\sigma,T]$ associated with the control $\mu\in\mathscr{R}$ starting from $u_{\tau,x,\nu}(\tau+\sigma),\dot u_{\tau,x,\nu}(\tau+\sigma)$, at time $\tau+\sigma$ in (13). Using the continuous dependence of the solution with respect to the state and the control, it is readily seen that the mapping $(\nu,\mu)\mapsto v_{\tau+\sigma,u_{\tau,x,\nu}(\tau+\sigma),\dot u_{\tau,x,\nu}(\tau+\sigma),\mu}(t)$ is continuous on $\mathscr{R}\times\mathscr{R}$ for each $t\in[\tau,T]$, namely, if ν^n stably converges to $\nu\in\mathscr{R}$ and μ^n stably converges to $\mu\in\mathscr{R}$, then $v_{\tau+\sigma,u_{\tau,x,\nu^n}(\tau+\sigma),\dot u_{\tau,x,\nu}(\tau+\sigma),\mu^n}$ pointwisely converges to $v_{\tau+\sigma,u_{\tau,x,\nu}(\tau+\sigma),\dot u_{\tau,x,\nu}(\tau+\sigma),\mu}$. By using the fiber product of Young measure (see Theorem 4.5 or [22, Theorem 3.3.1]), we deduce that

$$(\nu,\mu)\mapsto\int_{\tau+\sigma}^{T}\left[\int_Z J(t,v_{\tau+\sigma,u_{\tau,x,\nu}(\tau+\sigma),\dot u_{\tau,x,\nu}(\tau+\sigma)\mu}(t),z)\mu_t(dz)\right]dt$$

is continuous on $\mathscr{R}\times\mathscr{R}$. Consequently $\nu\mapsto V_J(\tau+\sigma,u_{\tau,x,\nu}(\tau+\sigma),\dot u_{\tau,x,\nu}(\tau+\sigma))$ is continuous on $\mathscr{R}$. Hence the mapping $\nu\mapsto\int_\tau^{\tau+\sigma}[\int_Z J(t,u_{\tau,x,\nu}(t),z)\nu_t(dz)]dt+V_J(\tau+\sigma,u_{\tau,x,\nu}(\tau+\sigma),\dot u_{\tau,x,\nu}(\tau+\sigma))$ is continuous on $\mathscr{R}$. By compactness of $\mathscr{R}$, there is a maximum point $\nu^1\in\mathscr{R}$ such that

$$W_J(\tau,x,y)=\int_\tau^{\tau+\sigma}\left[\int_Z J(t,u_{\tau,x,y,\nu^1}(t),z)\nu_t^1(dz)\right]dt+V_J(\tau+\sigma,u_{\tau,x,y,\nu^1}(\tau+\sigma)).$$

Similarly there is $\mu^2\in\mathscr{R}$ such that

$$\begin{aligned}&V_J(\tau+\sigma,u_{\tau,x,\nu^1}(\tau+\sigma),\dot u_{\tau,x,\nu^1}(\tau+\sigma))\\&\quad=\int_{\tau+\sigma}^{T}\left[\int_Z J(t,v_{\tau+\sigma,u_{\tau,x,\nu^1}(\tau+\sigma),\dot u_{\tau,x,\nu^1}(\tau+\sigma),\mu^2}(t),z)\mu_t^2(dz)\right]dt\end{aligned}$$

where

$$v_{\tau+\sigma,u_{\tau,x,\nu^1}(\tau+\sigma),\dot u_{\tau,x,\nu^1}(\tau+\sigma),\mu^2}(t)$$

denotes the trajectory solution associated with the control $\mu^2\in\mathscr{R}$ starting from $u_{\tau,x,\nu^1}(\tau+\sigma),\dot u_{\tau,x,\nu^1}(\tau+\sigma)$ at time $\tau+\sigma$ defined on $[\tau+\sigma,T]$

$$\begin{cases}0\in v_{\tau+\sigma,u_{\tau,x,y,\nu^1}(\tau+\sigma),\dot u_{\tau,x,y,\nu^1}(\tau+\sigma),\mu^2}(t)+A(t)v_{\tau+\sigma,u_{\tau,x,y\nu^1}(\tau+\sigma),\dot u_{\tau,x,y,\nu^1}(\tau+\sigma),\mu^2}(t),\\\quad+f(t,v_{\tau+\sigma,u_{\tau,x,y,\nu^1}(\tau+\sigma),\dot u_{\tau,x,y,\nu^1}(\tau+\sigma),\mu^2}(t))+\mathrm{bar}(\mu_t^2),\\v_{\tau+\sigma,u_{\tau,x,y,\nu^1}(\tau+\sigma),\dot u_{\tau,x,y,\nu^1}(\tau+\sigma),\mu^2}(\tau+\sigma)=u_{\tau,x,y,\nu^1}(\tau+\sigma),\\\dot v_{\tau+\sigma,u_{\tau,x,y,\nu^2}(\tau+\sigma),\dot u_{\tau,x,y,\nu^1}(\tau+\sigma),\mu^2}(\tau+\sigma)=\dot u_{\tau,x,y,\nu^1}(\tau+\sigma)\in D(A(\tau+\sigma)).\end{cases}\tag{14}$$

Let us set

$$\overline{\nu} := 1_{[\tau,\tau+\sigma]}\nu^1 + 1_{[\tau+\sigma,T]}\mu^2.$$

Then $\overline{\nu} \in \mathscr{R}$. Let $w_{\tau,x,y,\overline{\nu}}$ be the trajectory solution on $[\tau, T]$ associated with $\overline{\nu} \in \mathscr{R}$, that is,

$$\begin{cases} 0 \in \ddot{w}_{\tau,x,y,\overline{\nu}}(t) + A(t)\dot{w}_{\tau,x,y,\overline{\nu}}(t) + f(t, w_{\tau,x,y,\overline{\nu}}(t)) + \text{bar}(\overline{\nu}_t) \quad w_{\tau,x,y,\overline{\nu}}(\tau) = x \\ \dot{w}_{\tau,x,y,\overline{\nu}}(\tau) = y \in D(A(\tau)) \end{cases}$$

By uniqueness of the solution, we have

$$w_{\tau,x,y,\overline{\nu}}(t) = u_{\tau,x,y,\nu^1}(t), \ \forall t \in [\tau, \tau + \sigma],$$

$$w_{\tau,x,y,\overline{\nu}}(t) = v_{\tau+\sigma, u_{\tau,x,y,\nu^1}(\tau+\sigma), \dot{u}_{\tau,x,y,\nu^1}(\tau+\sigma), \mu^2}(t), \ \forall t \in [\tau + \sigma, T].$$

Coming back to the expression of V_J and W_J, we have

$$\begin{aligned} W_J(\tau, x, y) &= \int_\tau^{\tau+\sigma} \left[\int_Z J(t, u_{\tau,x,y,\nu^1}(t), z)\nu_t^1(dz) \right] dt \\ &\quad + \int_{\tau+\sigma}^T \left[\int_Z J(t, v_{\tau+\sigma, u_{\tau,x,\nu^1}(\tau+\sigma), \dot{u}_{\tau,x,\nu^1}(\tau+\sigma), \mu^2}(t), z)\mu_t^2(dz) \right] dt \\ &= \int_\tau^T \left[\int_Z J(t, w_{\tau,x,y,\overline{\nu}}(t), z)\overline{\nu}_t(dz) \right] dt \\ &\le \sup_{\nu \in \mathscr{R}} \left\{ \int_\tau^T \left[\int_Z J(t, u_{\tau,x,y,\nu}(t), z)\nu_t(dz) \right] dt \right\} = V_J(\tau, x, y). \end{aligned}$$

The proof is complete.

In the above evolution problem, we deal with second-order inclusion of the form

$$\begin{cases} 0 \in \ddot{u}_{x,y,\lambda}(t) + A(t)\dot{u}_{x,y,\lambda}(t) + f(t, u_{x,y,\lambda}(t)) + \text{bar}(\lambda_t), \ t \in [0, T] \\ u_{x,y,\lambda}(0) = x, \ \dot{u}_{x,\lambda}(0) = y \in D(A(0)) \end{cases}$$

with perturbed term f and $\text{bar}(\lambda_t)$. Now we focus to the evolution inclusion of the form

$$\begin{cases} 0 \in \dot{u}_{x,\lambda}(t) + A(t)u_{x,\lambda}(t) + f(t, u_{x,\lambda}(t)) + \text{bar}(\lambda_t), \ t \in [0, T] \\ u_{x,\lambda}(0) = x \in D(A(0)) \end{cases}$$

By Theorem 3.1, there is a unique Lipschitz solution $u_{x,\lambda}$ to this inclusion. Using the above techniques and Theorem 3.1, we have a result of dynamic principle that is similar to Theorem 4.9.

Theorem 4.10 (of dynamic programming principle) *Assume the hypothesis and notations of Theorem* 3.1*, and let $x \in E$, $\tau < T$ and $\sigma > 0$ such that $\tau + \sigma < T$. Assume that $J : [0, T] \times H \times Z \to \mathbb{R}$ is bounded and continuous. Let us consider the value function*

$$V_J(\tau, x, y) = \sup_{\nu \in \mathscr{R}} \int_\tau^T \left[\int_Z J(t, u_{\tau,x,\nu}(t), z)\nu_t(dz) \right] dt, \quad (\tau, x) \in [0, T] \times D(A(\tau))$$

where $u_{\tau,\nu}$ is the Lipschitz solution to the evolution inclusion defined on $[\tau, T]$ associated with the control $\nu \in \mathscr{R}$ starting from x, at time τ

$$\begin{cases} 0 \in \dot{u}_{\tau,x,\nu}(t) + A(t)u_{\tau,x,\nu}(t) + f(t, u_{\tau,x,\nu}(t)) + \mathrm{bar}(\nu_t), \\ u_{\tau,x,\nu}(\tau) = x \in D(A(\tau)). \end{cases}$$

Then the following holds:

$$V_J(\tau, x) = \sup_{\nu \in \mathscr{R}} \left\{ \int_\tau^{\tau+\sigma} \left[\int_Z J(t, u_{\tau,x,\nu}(t), z)\nu_t(dz) \right] dt + V_J(\tau+\sigma, u_{\tau,x,\nu}(\tau+\sigma)) \right\}$$

with

$$V_J(\tau + \sigma, u_{\tau,x,\nu}(\tau + \sigma)) = \sup_{\mu \in \mathscr{R}} \int_{\tau+\sigma}^T \left[\int_Z J(t, v_{\tau+\sigma,u_{\tau,x,\nu}(\tau+\sigma),\mu}(t), z)\mu_t(dz) \right] dt$$

where $v_{\tau+\sigma,u_{\tau,x,\nu}(\tau+\sigma),\mu}$[2] *is the Lipschitz solution defined on $[\tau + \sigma, T]$ associated with the control $\mu \in \mathscr{R}$ starting from $u_{\tau,x,\nu}(\tau + \sigma)$ at time $\tau + \sigma$*

$$\begin{cases} 0 \in \dot{v}_{\tau+\sigma,u_{\tau,x,\nu}(\tau+\sigma),\mu}(t) + A(t)v_{\tau+\sigma,u_{\tau,x,\nu}(\tau+\sigma)),\mu}(t) \\ \qquad\qquad + f(t, v_{\tau+\sigma,u_{\tau,x,y,\nu}(\tau+\sigma),\mu}(t)) + \mathrm{bar}(\mu_t), \\ v_{\tau+\sigma,u_{\tau,x,\nu}(\tau+\sigma),\mu}(\tau + \sigma) = u_{\tau,x,\nu}(\tau + \sigma) \in D(A(\tau + \sigma)). \end{cases}$$

Let us mention a useful lemma. See also [16, 22, 23] for related results.

Lemma 4.7 *Assume the hypothesis and notations of Theorem* 3.1*. Let Z be a compact subset in H, and $\mathscr{M}_+^1(Z)$ is endowed with the narrow topology and $\mathscr{R}$ the space of relaxed controls associated with Z. Let $\Lambda : [0, T] \times H \times \mathscr{M}_+^1(Z) \to \mathbb{R}$ be an upper semicontinuous function such that the restriction of Λ to $[0, T] \times B \times \mathscr{M}_+^1(Z)$ is bounded on any bounded subset B of H. Let $(t_0, x_0) \in [0, T] \times E$. If*

---

[2]It is necessary to write completely the expression of the trajectory $v_{\tau+\sigma,u_{\tau,x,\nu}(\tau+\sigma),\mu}$ that depends on $(\nu, \mu) \in \mathscr{R} \times \mathscr{R}$ in order to get the continuous dependence with respect to $\nu \in \mathscr{R}$ of $V_J(\tau + \sigma, u_{\tau,x,\nu}(\tau + \sigma))$.

$\max_{\mu \in \mathcal{M}_+^1(Z)} \Lambda(t_0, x_0, \mu) < -\eta < 0$ for some $\eta > 0$, then there exist $\sigma > 0$ such that

$$\sup_{\nu \in \mathcal{R}} \int_{t_0}^{t_0+\sigma} \Lambda(t, u_{t_0,x_0,\nu}(t), \nu_t)\, dt < -\frac{\sigma\eta}{2}$$

where $u_{t_0,x_0,\nu}$ is the trajectory solution associated with the control $\nu \in \mathcal{R}$ and starting from x_0 at time t_0

$$\begin{cases} 0 \in \dot{u}_{t_0,x_0,\nu}(t) + A(t)u_{t_0,x_0,\nu}(t) + f(t, u_{t_0,x_0,\nu}(t) + \mathrm{bar}(\nu_t),\ t \in [t_0, T], \\ u_{t_0,x_0,\nu}(t_0) = x_0 \in D(A(t_0)). \end{cases}$$

Proof By our assumption $\max_{\mu \in \mathcal{M}_+^1(Z)} \Lambda(t_0, x_0, \mu) < -\eta < 0$ for some $\eta > 0$. As the function $(t, x, \mu) \mapsto \Lambda(t, x, \mu)$ is upper semicontinuous, so is the function

$$(t, x) \mapsto \max_{\mu \in \mathcal{M}_+^1(Z)} \Lambda(t, x, \mu).$$

Hence there exists $\zeta > 0$ such that

$$\max_{\mu \in \mathcal{M}_+^1(Z)} \Lambda(t, x, \mu) < -\frac{\eta}{2}$$

for $0 < t - t_0 \leq \zeta$ and $||x - x_0|| \leq \zeta$. Thus, for small values of σ, we have

$$||u_{t_0,x_0,\nu}(t) - u_{t_0,x_0,\nu}(t_0)|| \leq \zeta$$

for all $t \in [t_0, t_0 + \sigma]$ and for all $\nu \in \mathcal{R}$ because $||\dot{u}_{t_0,x_0,\nu}(t)|| \leq K =$ Constant for all $\nu \in \mathcal{R}$ and for all $t \in [0, T]$ so that $||u_{t_0,x_0,\nu}(t)|| \leq L =$ Constant for all $\nu \in \mathcal{R}$ and for all $t \in [0, T]$ Hence $t \mapsto \Lambda(t, u_{t_0,x_0,\nu}(t), \nu_t)$ is bounded and Lebesgue-measurable on $[t_0, t_0 + \sigma]$. Then by integrating

$$\int_{t_0}^{t_0+\sigma} \Lambda(t, u_{t_0,x_0,\nu}(t), \nu_t)\, dt \leq \int_{t_0}^{t_0+\sigma} \left[\max_{\mu \in \mathcal{M}_+^1(Z)} \Lambda(t, u_{t_0,x_0,\nu}(t), \mu)\right] dt < -\frac{\sigma\eta}{2}.$$

The proof is complete.

Now to finish the paper, we provide a direct application to the viscosity solution to the evolution inclusion of the form

$$\begin{cases} 0 \in \dot{u}_{x,\lambda}(t) + A(t)u_{x,\lambda}(t) + f(t, u_{x,\lambda}(t)) + \mathrm{bar}(\lambda_t),\ t \in [0, T] \\ u_{x,\lambda}(0) = x \in D(A(0)) \end{cases}$$

where $A(t)$ is a convex weakly compact valued $H \to cwk(H)$ maximal monotone operator.

Theorem 4.11 *Let for every $t \in [0, T]$, $A(t) : H \to cwk(H)$ be a convex weakly compact valued maximal monotone operator satisfying*

$(H1)$ *there exists a real constant $\alpha \geq 0$ such that*

$$\mathrm{dis}(A(t), A(s)) \leq \alpha(t - s) \text{ for } 0 \leq s \leq t \leq T.$$

$(H2)$ *there exists a nonnegative real number c such that*

$$\|A^0(t, x)\| \leq c(1 + \|x\|), t \in [0, T], x \in H$$

$(H3)$ *$(t, x) \mapsto A(t)x$ is scalarly upper semicontinuous on $[0, T] \times H$.*

Let Z be a compact subset in H, and let $\mathscr{R}$ be the space of relaxed controls associated with Z. Let $f : [0, T] \times H \to H$ be a continuous mapping satisfying

(i) *there is $M > 0$ such that $||f(t, x)|| \leq M(1 + ||x||)$ for all (t, x) in $[0, T] \times H$,*
(ii) *$||f(t, x) - f(t, y)|| \leq M||x - y||$ for all $(t, x, y) \in [0, T] \times H \times H$.*

Assume that $J : [0, T] \times H \times Z \to \mathbb{R}$ is bounded and continuous. Let us consider the value function

$$V_J(\tau, x) = \sup_{\nu \in \mathscr{R}} \int_\tau^T \left[\int_Z J(t, u_{\tau,x,\nu}(t), z)\nu_t(dz) \right] dt, \ (\tau, x) \in [0, T] \times H$$

where $u_{\tau,x,\nu}$ is the trajectory solution on $[\tau, T]$ of the evolution inclusion associated with $A(t)$ and the control $\nu \in \mathscr{R}$ and starting from $x \in H$ at time τ

$$\begin{cases} 0 \in \dot{u}_{\tau,x,\nu}(t) + A(t, u_{\tau,x,\nu}(t)) + f(t, u_{\tau,x,\nu}(t)) + \mathrm{bar}(\nu_t), \ t \in [\tau, T] \\ u_{\tau,x,\nu}(\tau) = x \in H \end{cases}$$

and the Hamiltonian

$$\begin{aligned} &H(t, x, \rho) \\ &= \sup_{\mu \in \mathscr{M}_+^1(Z)} \left[-\langle \rho, \mathrm{bar}(\mu) \rangle + \int_Z J(t, x, z)\mu(dz) \right] + \delta^*(\rho, -f(t, x) - A(t, x)) \end{aligned}$$

where $(t, x, \rho) \in [0, T] \times H \times H$. Then, V_J is a viscosity subsolution of the HJB equation

$$\frac{\partial U}{\partial t}(t, x) + H(t, x, \nabla U(t, x)) = 0,$$ [3]

[3] Where ∇U is the gradient of U with respect to the second variable.

that is, for any $\varphi \in C^1([0,T]) \times H)$ *for which* $V_J - \varphi$ *reaches a local maximum at* $(t_0, x_0) \in [0,T] \times H$, *we have*

$$H(t_0, x_0, \nabla\varphi(t_0, x_0)) + \frac{\partial\varphi}{\partial t}(t_0, x_0) \geq 0.$$

Proof Assume by contradiction that there exists a $\varphi \in C^1([0,T] \times H)$ and a point $(t_0, x_0) \in [0,T] \times H$ for which

$$\frac{\partial\varphi}{\partial t}(t_0, x_0) + H(t_0, x_0, \nabla\varphi(t_0, x_0)) \leq -\eta < 0 \quad \text{for} \quad \eta > 0.$$

Applying Lemma 3.5 by taking

$$\Lambda(t, x, \mu) = -\langle \nabla\varphi(t, x), \mathrm{bar}(\mu) \rangle + \int_Z J(t, x, z)\mu(dz)$$
$$+ \delta^*(\nabla\varphi(t, x), -f(t, x) - A(t, x)) + \frac{\partial\varphi}{\partial t}(t, x)$$

yields some $\sigma > 0$ such that

$$\begin{aligned}
\sup_{\nu \in \mathscr{R}} \Bigg(\int_{t_0}^{t_0+\sigma} \left[\int_Z J(t, u_{t_0,x_0,\nu}(t), z)\nu_t(dz) \right] dt - \int_{t_0}^{t_0+\sigma} \langle \nabla\varphi(t, u_{t_0,x_0,\nu}(t)), \mathrm{bar}(\nu_t) \rangle \, dt \\
+ \int_{t_0}^{t_0+\sigma} \delta^*(\nabla\varphi(t, u_{t_0,x_0,\nu}(t)), -f(t, u_{t_0,x_0,\nu^n}(t)) - A(t, u_{t_0,x_0,\nu}(t))) \, dt \\
+ \int_{t_0}^{t_0+\sigma} \frac{\partial\varphi}{\partial t}(t, u_{t_0,x_0,\nu}(t)) \, dt \Bigg] \qquad (15) \\
\leq -\frac{\sigma\eta}{2}
\end{aligned}$$

where $u_{t_0,x_0,\nu}$ is the trajectory solution associated with the control $\nu \in \mathscr{R}$ starting from x_0 at time t_0

$$\begin{cases} 0 \in \dot{u}_{t_0,x_0,\nu}(t) + A(t, u_{t_0,x_0,\nu}(t)) + f(t, u_{t_0,x_0,\nu}(t)) + \mathrm{bar}(\nu_t), \;\; t \in [t_0, T] \\ u_{t_0,x_0,\nu}(t_0) = x_0. \end{cases}$$

Applying the dynamic programming principle (Theorem 4.10) gives

$$\begin{aligned}
V_J(t_0, x_0) = \sup_{\nu \in \mathscr{R}} \Bigg(\int_{t_0}^{t_0+\sigma} \left[\int_Z J(t, u_{t_0,x_0,\nu}(t), z)\nu_t(dz) \right] dt + V_J(t_0 \\
+\sigma, u_{t_0,x_0,\nu}(t_0 + \sigma)) \Bigg].
\end{aligned} \qquad (16)$$

Since $V_J - \varphi$ has a local maximum at (t_0, x_0), for small enough σ

$$V_J(t_0, x_0) - \varphi(t_0, x_0) \geq V_J(t_0 + \sigma, u_{t_0,x_0,\nu}(t_0 + \sigma)) - \varphi(t_0 + \sigma, u_{t_0,x_0,\nu}(t_0 + \sigma)) \tag{17}$$

for all $\nu \in \mathscr{R}$. By (16), for each $n \in \mathbb{N}$, there exists $\nu^n \in \mathscr{R}$ such that

$$V_J(t_0, x_0) \leq \int_{t_0}^{t_0+\sigma} \left[\int_Z J(t, u_{t_0,x_0,\nu^n}(t)), z)\nu_t^n(dz) \right] dt + V_J(t_0 + \sigma, u_{t_0,x_0,\nu^n}(t_0 + \sigma)) + \frac{1}{n}. \tag{18}$$

From (17) and (18), we deduce that

$$\begin{aligned} V_J(t_0 + \sigma, u_{t_0,x_0,\nu^n}(t_0 + \sigma)) - \varphi(t_0 + \sigma, u_{t_0,x_0,\nu^n}(t_0 + \sigma)) \\ \leq \int_{t_0}^{t_0+\sigma} \left[\int_Z J(t, u_{t_0,x_0,\nu^n}(t)), z)\nu_t^n(dz) \right] dt + \frac{1}{n} \\ - \varphi(t_0, x_0) + V_J(t_0 + \sigma, u_{t_0,x_0,\nu^n}(t_0 + \sigma)). \end{aligned}$$

Therefore we have

$$0 \leq \int_{t_0}^{t_0+\sigma} \left[\int_Z J(t, u_{t_0,x_0,\nu^n}(t)), z)\nu_t^n(dz) \right] dt + \varphi(t_0 + \sigma, u_{t_0,x_0,\nu^n}(t_0 + \sigma)) - \varphi(t_0, x_0) + \frac{1}{n}. \tag{19}$$

As $\varphi \in C^1([0, T] \times H)$, we have

$$\begin{aligned} &\varphi(t_0 + \sigma, u_{t_0,x_0,\nu^n}(t_0 + \sigma)) - \varphi(t_0, x_0) \\ &= \int_{t_0}^{t_0+\sigma} \langle \nabla\varphi(t, u_{t_0,x_0,\nu^n}(t)), \dot{u}_{t_0,x_0,\nu^n}(t)\rangle \, dt + \int_{t_0}^{t_0+\sigma} \frac{\partial\varphi}{\partial t}(t, u_{t_0,x_0,\nu^n}(t))\, dt. \end{aligned} \tag{20}$$

Since u_{t_0,x_0,ν^n} is the trajectory solution starting from x_0 at time t_0

$$\begin{cases} 0 \in \dot{u}_{t_0,x_0,\nu^n}(t) + A(t, u_{t_0,x_0,\nu^n}(t)) + f(t, u_{t_0,x_0,\nu^n}(t)) + \mathrm{bar}(\nu_t^n), \ t \in [t_0, T] \\ u_{t_0,x_0,\nu^n}(t_0) = x_0 \end{cases}$$

so that (20) yields the estimate

$$
\begin{aligned}
&\varphi(t_0+\sigma, u_{t_0,x_0,\nu^n}(t_0+\sigma)) - \varphi(t_0, x_0)\\
&= \int_{t_0}^{t_0+\sigma} \langle \nabla\varphi(t, u_{t_0,x_0,\nu^n}(t)), \dot{u}_{t_0,x_0,\nu^n}(t)\rangle\, dt + \int_{t_0}^{t_0+\sigma} \frac{\partial\varphi}{\partial t}(t, u_{t_0,x_0,\nu^n}(t))\, dt\\
&\le -\int_{t_0}^{t_0+\sigma} \langle \nabla\varphi(t, u_{t_0,x_0,\nu^n}(t)), \mathrm{bar}(\nu_t^n) + f(t, u_{t_0,x_0,\nu^n}(t))\rangle\, dt\\
&\qquad + \int_{t_0}^{t_0+\sigma} \delta^*(\nabla\varphi(t, u_{t_0,x_0,\nu^n}(t)), -A(t, u_{t_0,x_0,\nu^n}(t)))\, dt\\
&\qquad\qquad + \int_{t_0}^{t_0+\sigma} \frac{\partial\varphi}{\partial t}(t, u_{t_0,x_0,\nu^n}(t))\, dt.
\end{aligned}
\tag{21}
$$

Inserting the estimate (21) into (19), we get

$$
\begin{aligned}
0 &\le \int_{t_0}^{t_0+\sigma} \left[\int_Z J(t, u_{t_0,x_0,\nu^n}(t)), z)\nu_t^n(dz)\right] dt \qquad (22)\\
&- \int_{t_0}^{t_0+\sigma} \langle \nabla\varphi(t, u_{t_0,x_0,\nu^n}(t)), \mathrm{bar}(\nu_t^n) + f(t, u_{t_0,x_0,\nu^n}(t))\rangle\, dt\\
&+ \int_{t_0}^{t_0+\sigma} \delta^*(\nabla\varphi(t, u_{t_0,x_0,\nu^n}(t)), -A(t, u_{t_0,x_0,\nu^n}(t))\, dt\\
&+ \int_{t_0}^{t_0+\sigma} \frac{\partial\varphi}{\partial t}(t, u_{t_0,x_0,\nu^n}(t))\, dt + \frac{1}{n}.
\end{aligned}
$$

Then (15) and (22) yield $0 \le -\frac{\sigma\eta}{2} + \frac{1}{n}$ for all $n \in \mathbb{N}$. By passing to the limit when n goes to ∞ in this inequality, we get a contradiction: $0 \le -\frac{\sigma\eta}{2}$. The proof is therefore complete.

Existence results for evolution inclusion driven by time-dependent maximal monotone operators $A(t)$ with single-valued perturbation f or convex weakly compact valued perturbation F of the form

$$-\dot{u}(t) \in A(t)u(t) + f(t, u(t))$$

or

$$-\dot{u}(t) \in A(t)u(t) + F(t, u(t))$$

are developed in [7, 8], while existence results for convex or nonconvex sweeping process in the form

$$-\dot{u}(t) \in N_{C(t)}(u(t)) + f(t, u(t))$$

or

$$-\dot{u}(t) \in N_{C(t)}(u(t)) + F(t, u(t))$$

where $C(t)$ is a closed convex (or nonconvex) moving set and $N_{C(t)}(u(t))$ is the normal cone of $C(t)$ at the point $u(t)$ is much studied so that our tools developed above allow to treat some further variants on the viscosity solution dealing with some specific maximal monotone operators $A(t)$ or convex or nonconvex sweeping process such as

$$\begin{cases} 0 \in \dot{u}_{t_0,x_0,\nu}(t) + N_{C(t)}(u_{t_0,x_0,\nu}(t)) + f(t, u_{\tau,x,\nu}(t)) + \text{bar}(\nu_t),\ t \in [t_0, T] \\ u_{t_0,x_0,\nu}(t_0) = x_0 \end{cases}$$

using the subdifferential of the distance function $d_{C(t)}x$.

We end the paper with some variational limit results which can be applied to further convergence problems in state-dependent convex sweeping process or second-order state-dependent convex sweeping process. See [1, 3, 34] and the references therein.

Theorem 4.12 *Let $C_n : [0, T] \to H$ and $C : [0, T] \to H$ be a sequence of convex weakly compact valued scalarly measurable bounded mappings satisfying*

(i) $\sup_n \sup_{t\in[0,T]} \mathcal{H}\left(C_n(t), C(t)\right) \le M < \infty$,
(ii) $\lim_n \mathcal{H}\left(C_n(t), C(t)\right) = 0$, *for each* $t \in [0, T]$.

Let (v_n) be a uniformly integrable sequence in $L^1_H([0, T])$ such that v_n converges for $\sigma(L^1_H([0, T]), L^\infty_H([0, T])$ to $v \in L^1_H([0, T])$, and let (u_n) be a uniformly bounded sequence $L^\infty_H([0, T])$ which pointwisely converges to $u \in L^\infty_H([0, T])$. Assume that $-v_n(t) \in N_{C_n(t)}(u_n(t))$ a.e., then

$$u(t) \in C(t) \text{ a.e. and } -v(t) \in N_{C(t)}(u(t)) \text{ a.e.}$$

Proof For simplicity, let $\rho_n(t) = \mathcal{H}\left(C_n(t), C(t)\right)$ for each $t \in [0, T]$. Firstly it is clear that the mappings ρ_n, $t \mapsto \delta^*(-v_n(t), C_n(t))$, $t \mapsto \delta^*(-v_n(t), C(t))$, and $t \mapsto \delta^*(-v(t), C(t))$ are measurable on $[0, T]$ and integrable by boundedness. By the Hormander formula for convex weakly compact set (see [19]), we have

$$|\delta^*(-v_n(t), C_n(t)) - \delta^*(-v_n(t), C(t))| \le ||v_n(t)||\rho_n(t)$$

so that

$$\delta^*(-v_n(t), C_n(t)) - \delta^*(-v_n(t), C(t)) \ge -||v_n(t)||\rho_n(t).$$

By $-v_n(t) \in N_{C_n(t)}(u_n(t))$, we have $\delta^*(-v_n(t), C_n(t)) + \langle v_n(t), u_n(t)\rangle \le 0$ so we get the estimation

$$-||v_n(t)||\rho_n(t) + \delta^*(-v_n(t), C(t)) + \langle v_n(t, u_n(t)\rangle \le 0$$

or

$$\delta^*(-v_n(t), C(t)) + \langle v_n(t), u_n(t)\rangle \leq ||v_n(t)||\rho_n(t).$$

Note that the mappings $t \mapsto \delta^*(-v_n(t), C(t)) + \langle v_n(t), u_n(t)\rangle$, and $t \mapsto ||v_n(t)||\rho_n(t)$ are integrable on $[0, T]$. Let B a measurable set in $[0, T]$ and then by integrating

$$\int_B \delta^*(-v_n(t), C(t))dt + \int_B \langle v_n(t), u_n(t)\rangle dt \leq \int_B ||v_n(t)||\rho_n(t)dt.$$

We note that the convex integrand $H(t, e) = \delta^*(e, C(t))$ defined on $[0, T] \times H$ is normal because $t \mapsto H(t, e)$ is continuous on $[0, T]$ and $e \mapsto H(t, e)$ is convex continuous on H, with $H(t, e) \geq \langle e, u(t)\rangle$ for all $(t, e) \in [0, T] \times H$. Consequently $H(t, -v_n(t)) = \delta^*(-v_n(t), C(t)) \geq \langle -v_n(t), u(t)\rangle$. But $(\langle -v_n, u\rangle)$ is uniformly integrable in $L^1_{\mathbb{R}}([0, T], dt)$, so that by virtue of the lower semicontinuity of the integral convex functional [22, Theorem 8.1.16], we have

$$\liminf_{n\to\infty} \int_B \delta^*(-v_n(t), C(t))dt \geq \int_B \delta^*(-v(t), C(t))dt. \tag{23}$$

Note that the sequence $\big(u_n(.) - u(.)\big)$ is uniformly bounded and pointwisely converges to 0, so that it converges to 0 uniformly on any uniformly integrable subset of $L^1_H([0, T], dt)$; in other terms, it converges to 0 with respect to the Mackey topology $\tau(L^\infty_H([0, T], dt), L^1_H([0, T], dt))$ (see [15]),[4] so that

$$\lim_{n\to\infty} \int_B \langle v_n(t), u_n(t) - u(t)\rangle dt = 0$$

because (v_n) is uniformly integrable. Consequently

$$\begin{aligned}\lim_{n\to\infty} \int_B \langle v_n(t), u_n(t)\rangle dt &= \lim_{n\to\infty} \int_B \langle v_n(t), u_n(t) - u(t)\rangle dt + \lim_{n\to\infty} \int_B \langle v_n(t), u(t)\rangle dt \\ &= \lim_{n\to\infty} \int_B \langle v_n(t), u(t)\rangle dt = \int_B \langle \dot{v}(t), u(t)\rangle dt. \end{aligned} \tag{24}$$

By our assumptions, $\rho_n(t)$ is bounded measurable and pointwisely converges to 0 and $||v_n(t)||$ is uniformly integrable; then similarly we have

$$\lim_n \int_B ||v_n(t)||\rho_n(t)dt = 0. \tag{25}$$

[4]If $H = \mathbb{R}^d$, one may invoke a classical fact that on bounded subsets of L^∞_H the topology of convergence in measure coincides with the topology of uniform convergence on uniformly integrable sets, i.e. on relatively weakly compact subsets, alias the Mackey topology. This is a lemma due to Grothendieck [33, Ch.5 §4 no 1 Prop. 1 and exercice].

Finally by passing to the limit when n goes to ∞ in

$$\int_B \delta^*(-v_n(t), C(t))dt + \int_B \langle v_n(t), u_n(t)\rangle dt \le \int_B ||v_n(t)||\rho_n(t)dt$$

and taking into account the above convergence limits (23), (24), and (25), we get

$$\int_B \delta^*(-v(t), C(t))dt + \int_B \langle v(t), u(t)\rangle dt \le 0.$$

As the function $t \mapsto \delta^*(-v(t), C(t)) + \langle v(t), u(t)$ is integrable on $[0, T]$ and this holds for every B measurable set in $[0, T]$, we get

$$\delta^*(-v(t), C(t))) + \langle v(t), u(t)\rangle \le 0 \text{ a.e.}$$

Furthermore, it is not difficult to check that $u(t) \in C(t)$ a.e. using (ii) and the fact that $u_n(t) \in C_n(t)$ for all $n \in \mathbb{N}$ and a.e. $t \in [0, T]$; therefore, we conclude that $-v(t) \in N_{C(t)}(u(t))$ a.e. The proof is complete.

Our tools allow to treat the variational limits for further evolution variational inequalities such as

Proposition 4.2 *Let $C_n : [0, T] \to H$ and $C : [0, T] \rightrightarrows H$ be a sequence of convex weakly valued scalarly measurable bounded mappings satisfying*

(i) $\sup_n \sup_{t\in[0,T]} \mathcal{H}\left(C_n(t), C(t)\right) \le M < \infty$,
(ii) $\lim_n \mathcal{H}\left(C_n(t), C(t)\right) = 0$, *for each $t \in [0, T]$.*

Let $B : H \to H$ be a linear continuous operator such that $\langle Bx, x\rangle > 0$ for all $x \in H \setminus \{0\}$. Let (v_n) be a uniformly bounded sequence in $L^\infty_H([0, T])$ such that v_n $\sigma(L^\infty_H([0, T]), L^1_H([0, T])$ converges to $v \in L^\infty_H([0, T])$, and let (u_n) be a uniformly bounded sequence $L^\infty_H([0, T])$ which pointwisely converges to $u \in L^\infty_H([0, T])$. Assume that $-v_n(t) \in N_{C_n(t)}(u_n(t)+Bv_n(t))$ for all $n \in \mathbb{N}$ and for a.e. $t \in [0, T]$. Then

$$u(t) + Bv(t) \in C(t) \text{ a.e. and } - v(t) \in N_{C(t)}(u(t) + Bv(t)) \text{ a.e.}$$

Proof Apply the notations of the proof of Theorem 4.12. Let $\rho_n(t) = \mathcal{H}\left(C_n(t), C(t)\right)$ for each $t \in [0, T]$. It is clear that the mappings ρ_n, $t \mapsto \delta^*(-v_n(t), C_n(t))$, $t \mapsto \delta^*(-v_n(t), C(t))$, and $t \mapsto \delta^*(-v(t), C(t))$ are measurable and integrable on $[0, T]$. By the Hormander formula for convex weakly compact sets (see [19]), we have

$$|\delta^*(-v_n(t), C_n(t)) - \delta^*(-v_n(t), C(t))| \le ||v_n(t)||\rho_n(t)$$

so that

$$\delta^*(-v_n(t), C_n(t)) - \delta^*(-v_n(t), C(t)) \ge -||v_n(t)||\rho_n(t).$$

By $-v_n(t) \in N_{C_n(t)}(u_n(t) + Bv_n(t))$, we have

$$\delta^*(-v_n(t), C_n(t)) + \langle v_n(t), u_n(t) + Bv_n(t)\rangle \leq 0.$$

Whence

$$\delta^*(-v_n(t), C(t)) + \langle v_n(t), u_n(t) + Bv_n(t)\rangle \leq ||v_n(t)||\rho_n(t)$$

Note that the mappings $t \mapsto \delta^*(-v_n(t), C(t)) + \langle v_n(t), u_n(t) + Bv_n(t)\rangle$, and $t \mapsto ||v_n(t)||\rho_n(t)$ are integrable on $[0, T]$ so that by integrating on any measurable set $L \subset [0, T]$

$$\int_L \delta^*(-v_n(t), C(t))dt + \int_L \langle v_n(t), u_n(t)\rangle dt + \int_L \langle v_n(t), Bv_n(t)\rangle dt$$
$$\leq \int_L ||v_n(t)||\rho_n(t)dt.$$

Since (v_n) $\sigma(L^\infty_H([0, T]), L^1_H([0, T])$ converges to $v \in L^\infty_H([0, T])$, it is not difficult to check that (Bv_n) converges for $\sigma(L^\infty_H([0, T]), L^1_H([0, T])$ to $Bv \in L^1_H([0, T])$, arguing as in [11, Theorem 4.1]. As a consequence, the sequence $(u_n + Bv_n)$ converges for $\sigma(L^\infty_H([0, T]), L^1_H([0, T])$ to $u + Bv \in L^\infty_H([0, T])$. From $u_n(t) + Bv_n(t) \in C_n(t)$, we deduce

$$\int_L \langle e, u_n(t) + Bv_n(t)\rangle dt \leq \int_L \delta^*(e, C_n(t))dt$$

for every $e \in H$ and for every measurable set $L \subset [0, T]$. By passing to the limit in this inequality, we get

$$\int_L \langle e, u(t) + Bv(t)\rangle dt \leq \limsup_n \int_L \delta^*(e, C_n(t))dt \leq \int_L \delta^*(e, C(t))dt.$$

It follows that

$$\langle e, u(t) + Bv(t)\rangle \leq \delta^*(e, C(t)) \text{ a.e.}$$

By [19, Proposition III.35], we deduce that $u(t) + Bv(t) \in C(t)$ a.e. As in Theorem 3.1, we have already stated that for every measurable set $L \subset [0, T]$,

$$\lim_n \int_L \langle u_n(t), v_n(t)\rangle dt = \int_L \langle u(t), v(t)\rangle dt, \tag{26}$$

$$\lim_n \int_L |v_n(t)||\rho_n(t)dt = 0, \tag{27}$$

$$\liminf_n \int_B \delta^*(-v(t), C_n(t))dt \geq \int_B \delta^*(-v(t), C_n(t))dt. \tag{28}$$

Now set $\varphi(x) = \langle x, Bx\rangle$ for all $x \in H$. Then $\varphi(x)$ is a nonnegative lower semicontinuous and convex function defined on H. So we have

$$\int_L \langle v_n(t), Bv_n(t)\rangle dt = \int_L \varphi(v_n(t))dt.$$

By lower semicontinuity of convex integral functional [19, 22, 23], we get

$$\liminf_n \int_L \langle v_n(t), Bv_n(t)\rangle dt$$
$$= \liminf_n \int_L \varphi(v_n(t))dt \geq \int_L \varphi(v(t))dt = \int_L \langle v(t), Bv(t)\rangle dt.$$

Taking into consideration the above stated limits (26), (27), (28) and passing to the limit when n goes to ∞ in the inequality

$$\int_L \delta^*(-v_n(t), C(t))dt + \int_L \langle v_n(t), u_n(t)\rangle dt + \int_L \langle v_n(t), Bv_n(t)\rangle dt$$
$$\leq \int_L ||v_n(t)||\rho_n(t)dt,$$

we get

$$\int_L \delta^*(-v(t), C(t))dt + \int_L \langle v(t), u(t) + Bv(t)\rangle dt \leq 0$$

for every measurable set $L \subset [0, T]$. Since the mapping $t \mapsto \delta^*(-v(t), C(t)) + \langle v(t), u(t) + Bv(t)\rangle$ is integrable on $[0, T]$, we have

$$\delta^*(-v(t), C(t)) + \langle v(t), u(t) + Bv(t)\rangle \leq 0 \text{ a.e.}$$

As $u(t) + Bv(t) \in C(t)$ a.e., this yields $-v(t) \in N_{C(t)}(u(t) + Bv(t))$ a.e. The proof is complete.

References

1. Adly S, Haddad T (2018) An implicit sweeping process approach to quasistatic evolution variational inequalities. Siam J Math Anal 50(1):761–778
2. Adly S, Haddad T, Thibault L (2014) Convex sweeping process in the framework of measure differential inclusions and evolution variational inequalities. Math Program 148(1–2, Ser. B):5–47
3. Aliouane F, Azzam-Laouir D, Castaing C, Monteiro Marques MDP (2018, Preprint) Second order time and state dependent sweeping process in Hilbert space

4. Attouch H, Cabot A, Redont P (2002) The dynamics of elastic shocks via epigraphical regularization of a differential inclusion. Barrier and penalty approximations. Adv Math Sci Appl 12(1):273–306. Gakkotosho, Tokyo
5. Azzam-Laouir D, Izza S, Thibault L (2014) Mixed semicontinuous perturbation of nonconvex state-dependent sweeping process. Set Valued Var Anal 22:271–283
6. Azzam-Laouir D, Makhlouf M, Thibault L (2016) On perturbed sweeping process. Appl Anal 95(2):303–322
7. Azzam-Laouir D, Castaing C, Monteiro Marques MDP (2017) Perturbed evolution problems with continuous bounded variation in time and applications. Set-Valued Var Anal. https://doi.org/10.1007/s11228-017-0432-9
8. Azzam-Laouir D, Castaing C, Belhoula W, Monteiro Marques MDP (2017, Preprint) Perturbed evolution problems with absolutely continuous variation in time and applications
9. Barbu (1976) Nonlinear semigroups and differential equations in Banach spaces. Noordhoff International Publisher, Leyden
10. Benabdellah H, Castaing C (1995) BV solutions of multivalued differential equations on closed moving sets in Banach spaces. Banach center publications, vol 32. Institute of Mathematics, Polish Academy of Sciences, Warszawa
11. Benabdellah H, Castaing C, Salvadori A (1997) Compactness and discretization methods for differential inclusions and evolution problems. Atti Sem Mat Fis Univ Modena XLV:9–51
12. Brezis H (1972) Opérateurs maximaux monotones dans les espaces de Hilbert et equations dévolution. Lectures notes 5. North Holland Publishing Co, Amsterdam
13. Brezis H (1979) Opérateurs maximaux monotones et semi-groupes de contraction dans un espace de Hilbert. North Holland Publishing Co, Amsterdam
14. Castaing C (1970) Quelques résultats de compacité liés a l' intégration. C R Acd Sci Paris 270:1732–1735; et Bulletin Soc Math France 31:73–81 (1972)
15. Castaing C (1980) Topologie de la convergence uniforme sur les parties uniformément intégrables de L^1_E et théorèmes de compacité faible dans certains espaces du type Köthe-Orlicz. Travaux Sém Anal Convexe 10(1):27. exp. no. 5
16. Castaing C, Marcellin S (2007) Evolution inclusions with pln functions and application to viscosity and control. JNCA 8(2):227–255
17. Castaing C, Monteiro Marques MDP (1996) Evolution problems associated with nonconvex closed moving sets with bounded variation. Portugaliae Mathematica 53(1):73–87; Fasc
18. Castaing C, Monteiro Marques MDP (1995) BV Periodic solutions of an evolution problem associated with continuous convex sets. Set Valued Anal 3:381–399
19. Castaing C, Valadier M (1977) Convex analysis and measurable multifunctions. Lectures notes in mathematics. Springer, Berlin, p 580
20. Castaing C, Duc Ha TX, Valadier M (1993) Evolution equations governed by the sweeping process. Set-Valued Anal 1:109–139
21. Castaing C, Salvadori A, Thibault L (2001) Functional evolution equations governed by nonconvex sweeeping process. J Nonlinear Convex Anal 2(2):217–241
22. Castaing C, Raynaud de Fitte P, Valadier M (2004) Young measures on topological spaces with applications in control theory and probability theory. Kluwer Academic Publishers, Dordrecht
23. Castaing C, Raynaud de Fitte P, Salvadori A (2006) Some variational convergence results with application to evolution inclusions. Adv Math Econ 8:33–73
24. Castaing C, Ibrahim AG, Yarou M (2009) Some contributions to nonconvex sweeping process. J Nonlinear Convex Anal 10(1):1–20
25. Castaing C, Monteiro Marques MDP, Raynaud de Fitte P (2014) Some problems in optimal control governed by the sweeping process. J Nonlinear Convex Anal 15(5):1043–1070
26. Castaing C, Monteiro Marques MDP, Raynaud de Fitte P (2016) A Skorohod problem governed by a closed convex moving set. J Convex Anal 23(2):387–423
27. Castaing C, Le Xuan T, Raynaud de Fitte P, Salvadori A (2017) Some problems in second order evolution inclusions with boundary condition: a variational approach. Adv Math Econ 21:1–46
28. Colombo G, Goncharov VV (1999) The sweeping processes without convexity. Set Valued Anal 7:357–374

29. Cornet B (1983) Existence of slow solutions for a class of differential inclusions. J Math Anal Appl 96:130–147
30. Edmond JF, Thibault L (2005) Relaxation and optimal control problem involving a perturbed sweeping process. Math Program Ser B 104:347–373
31. Flam S, Hiriart-Urruty J-B, Jourani A (2009) Feasibility in finite time. J Dyn Control Syst 15:537–555
32. Florescu LC, Godet-Thobie C (2012) Young measures and compactness in measure spaces. De Gruyter, Berlin
33. Grothendieck A (1964) Espaces vectoriels topologiques Mat, 3rd edn. Sociedade de matematica, Saõ Paulo
34. Haddad T, Noel J, Thibault L (2016) Perturbed Sweeping process with subsmooth set depending on the state. Linear Nonlinear Anal 2(1):155–174
35. Henry C (1973) An existence theorem for a class of differential equations with multivalued right-hand side. J Math Anal Appl 41:179–186
36. Idzik A (1988) Almost fixed points theorems. Proc Am Math Soc 104:779–784
37. Kenmochi N (1981) Solvability of nonlinear evolution equations with time-dependent constraints and applications. Bull Fac Educ Chiba Univ 30:1–87
38. Kunze M, Monteiro Marques MDP (1997) BV solutions to evolution problems with time-dependent domains. Set Valued Anal 5:57–72
39. Monteiro Marques MDP (1984) Perturbations convexes semi-continues supérieurement de problèmes d'évolution dans les espaces de Hilbert, vol 14. Séminaire d'Analyse Convexe, Montpellier, exposé n 2
40. Monteiro Marques MDP (1993) Differential inclusions nonsmooth mechanical problems, shocks and dry friction. Progress in nonlinear differential equations and their applications, vol 9. Birkhauser, Basel
41. Moreau JJ (1977) Evolution problem associated with a moving convex set in a Hilbert Space. J Differ Equ 26:347–374
42. Moreau JJ, Valadier M (1987) A chain rule involving vector functions of bounded variations. J Funct Anal 74(2):333–345
43. Paoli L (2005) An existence result for non-smooth vibro-impact problem. J Differ Equ 211(2):247–281
44. Park S (2006) Fixed points of approximable or Kakutani maps. J Nonlinear Convex Anal 7(1):1–17
45. Recupero V (2016) Sweeping processes and rate independence. J Convex Anal 23:921–946
46. Rockafellar RT (1971) Integrals which are convex functionals, II. Pac J Math 39(2):439–369
47. Saidi S, Thibault L, Yarou M (2013) Relaxation of optimal control problems involving time dependent subdifferential operators. Numer Funct Anal Optim 34(10):1156–1186
48. Schatzman M (1979) Problèmes unilatéraux d' évolution du second ordre en temps. Thèse de Doctorat d' Etates Sciences Mathématiques, Université Pierre et Marie Curie, Paris 6
49. Thibault L (1976) Propriétés des sous-différentiels de fonctions localement Lipschitziennes définies sur un espace de Banach séparable. Applications. Thèse, Université Montpellier II
50. Thibault L (2003) Sweeping process with regular and nonregular sets. J Differ Equ 193:1–26
51. Valadier M (1988) Quelques résultats de base concernant le processus de la rafle. Sém. Anal. Convexe, Montpellier, vol 3
52. Valadier M (1990) Lipschitz approximations of the sweeping process (or Moreau) process. J Differ Equ 88(2):248–264
53. Vladimirov AA (1991) Nonstationary dissipative evolution equation in Hilbert space. Nonlinear Anal 17:499–518
54. Vrabie IL (1987) Compactness methods for Nonlinear evolutions equations. Pitman monographs and surveys in pure and applied mathematics, vol 32. Longman Scientific and Technical, Wiley/New York

Plausible Equilibria and Backward Payoff-Keeping Behavior

Yuhki Hosoya

Abstract This paper addresses Selten's chain store paradox. We view this paradox as the phenomenon whereby the subgame perfect equilibria (SPEs) of some games are not credible. To solve this problem, we construct a refinement of Nash equilibria (NEs) called plausible equilibria. If an NE is included in this refinement, then the chain store paradox phenomenon does not occur and this equilibrium is credible. This paper analyzes the properties of this refinement and presents two results. First, every SPE of a zero-sum game with perfect information is plausible. Second, the notion of plausibility removes a bad equilibrium of the coordination game.

Keywords Chain store paradox · Plausible equilibria · Maximin criteria · Backward payoff-keeping behavior · Degree of implausibility

Article type: Research Article
Received: October 4, 2017
Revised: Feburary 1, 2018

1 Introduction

This paper describes a solution to Selten's [8] chain store paradox. Our approach differs from the standard approach. We view this paradox as the phenomena whereby the subgame perfect equilibria (SPEs) of some games are not credible,

JEL Codes: C72, C79
Mathematics Subject Classification (2010): 91A10, 91A18, 91A20

Y. Hosoya (✉)
Department of Economics, Kanto-Gakuin University, Fuchu-shi, Tokyo, Japan
e-mail: hosoya@kanto-gakuin.ac.jp

S. Kusuoka, T. Maruyama (eds.), *Advances in Mathematical Economics*, Advances in Mathematical Economics 22, https://doi.org/10.1007/978-981-13-0605-1_3

and some Nash equilibria (NEs) are credible even though they are not SPE. This situation can be solved using a refinement of NE that avoids this paradox. In this paper, we define such a refinement and demonstrate some of its properties.

We stress that the problem of the chain store paradox still remains, though many people say that it was solved by Kreps and Wilson [2]. Their study connects this paradox to the problem of reputation,[1] and solves this problem using the notion of sequential equilibria.[2] Most articles and books referring to this paradox mention this paper, and thus this paradox is treated as the problem of reputation.[3]

However, the interpretation of the paradox by Kreps and Wilson [2] is different from that by Selten [8]. More precisely, Kreps and Wilson [2] interpret this paradox as the inconsistency between the actual behavior of firms reported in Scherer [7] and the SPE of long-term repetitions of the entry deterrence game. This problem is different from the problem considered by Selten [8]. Selten stated that this problem is "an inconsistency between game theoretical reasoning and plausible human behavior," and "plausible human behavior" is not the same as the "actual behavior of firms" in Kreps-Wilson, because Selten's "behavior" is **behavior of a human that is confronted with an abstract game**. Selten's motivation is devoted to the abstract game itself, and thus he cannot change the analyzed game from the simple entry deterrence game to some other game. In contrast, in Kreps-Wilson, the "actual behavior of firms" means behavior of firms in the **real world**, and thus they can modify the game to introduce some imperfect information structure for explaining this phenomenon.[4]

We now try to explain the original chain store paradox of Selten [8]. Selten first constructs two games and demonstrates that most people probably opt against the strategy of unique SPE in these games. The reason why this strategy is denied is because there is an alternative strategy that seems to gain a larger payoff. Although

---

[1]In the same year, Milgrom and Roberts [5] and Kreps et al. [4] also treated the problem of reputation.

[2]See Kreps and Wilson [3] for a detailed argument.

[3]For example, see Fudenberg and Tirole [1] or Osborne and Rubinstein [6].

[4]In section 2 of Selten [8], he wrote the following arguments. "As we shall see in section 8, only the induction theory is game theoretically correct. Logically, the induction argument cannot be restricted to the last periods of the game. There is no way to avoid the conclusion that it applies to all periods of the game. Nevertheless the deterrence theory is much more convincing. If I had to play the game in the role of player A, I would follow the deterrence theory. I would be very surprised if it failed to work. From my discussions with friends and colleagues, I get the impression that most people share this inclination. In fact, up to now I met nobody who said that he would behave according to the induction theory. My experience suggests that mathematically trained persons recognize the logical validity of the induction argument, but they refuse to accept it as a guide to practical behavior." The ground for our claim that Selten's view of this paradox is a conflict between theory and experiments is the above paragraphs. Therefore, Selten cannot solve this paradox by modifying the game. However, Kreps and Wilson introduce a (potential) existence of tough player and solve this problem. The existence of tough player means that they modify the game from Selten's chain store game, and thus this approach is not the solution of the original chain store paradox considered by Selten. In contrast, the notion of DIP (we will define this term later) can reflect Selten's original idea to some extent.

this alternative strategy is not the best response for the strategy profile of unique SPE, it seems to be more profitable than the strategy of SPE, since if he/she chooses this strategy, then other players will probably change their strategy in response.

It is useful to consider the first type of chain store game of Selten [8]. This game has 21 players and consists of 20 stages. In the i-th stage, player i chooses either E or N. If player i chooses N, then he/she gains payoff 1 and player 0 gains payoff 5. Otherwise, player 0 chooses either A or C. If he/she chooses A, then both players i and 0 gain payoff 0. If he/she chooses C, then both players i and 0 gain payoff 2.

The unique SPE of this game is as follows: player 0 always chooses C , and every other player chooses E. Selten [8] calls this strategy the inductive theory. Meanwhile, Selten [8] introduces another strategy, called the deterrence theory. In deterrence theory, player 0 chooses A until stage 17 and chooses C in stages 18, 19, and 20.

If SPE arises, then player 0 gains 40. Suppose that player 0 chooses to follow the deterrence theory. If no other player deviates from the SPE strategy, then the payoff of player 0 is only 6. However, this scenario is not realistic, since late players can predict that player 0 will choose A, and if so, N is more profitable than E. If more than seven people change their choice, then the payoff for player 0 is more than 41. Hence, the deterrence theory seems to be more profitable than the inductive theory. This is the original chain store paradox.

We simplify this paradox as follows:

1. Suppose that some NE s^* tends to be realized.
2. Player i intentionally changes his/her action to something different from that in s^*.
3. Other players observe this action and change their action according to "some criterion."
4. Player i gains a payoff more than that in s^*. Thus, player i rejects strategy s_i^*.
5. s^* is not realized.

Thus, we want to define a refinement of NE in which these phenomena do not occur. To do this, we must clarify the notion of "some criterion" above. In the chain store paradox, players 1-20 mispredict the choice of player 0 and therefore change their action to a more sound choice. This "sound choice" seems to be the **maximin behavior**. Therefore, **we assume that this criterion is the maximin criterion** and thus define a refinement called the **plausible equilibrium**. Roughly speaking, plausible equilibria are NEs where no player can gain when he/she changes his/her action to control other players' actions.

Together with the above notion, we define the **degree of implausibility** (DIP) of NE. In our definition, an NE is plausible if and only if the DIP of this NE is zero. The reason why we consider DIP is that the plausibility is too strict and there might be no plausible equilibrium in many games. For example, Selten's chain store game has no plausible equilibrium. In contrast, there is an NE whose DIP is 1 in Selten's game, which seems to correspond with the deterrence theory. The DIP of the unique SPE in Selten's game is 12, which is much higher than in the above NE. We think that this is an explanation for Selten's original chain store paradox.

Using the notion of plausible equilibria, we can interpret likely behavior in an ordinal coordination game. That is, the good equilibrium is a unique plausible equilibrium of a long-term repeated game. We can interpret this result as follows: if the bad equilibrium tends to occur, player 1 denies this action and chooses the strategy of the good equilibrium. Then, after the next period, player 2 changes his action to the same strategy. Although player 1 loses his/her payoff in the first period, he/she eventually gains if the game is sufficiently long. Hence, the bad equilibrium is not plausible. Meanwhile, player 1 can change his/her action even when the good equilibrium tends to occur. However, this change cannot improve his/her long-run payoff. Therefore, the good equilibrium remains plausible.

In Sect. 2, we provide a rigorous definition of plausible equilibria and analyze some properties of this solution. In Sect. 3, we introduce several applications where plausible equilibria have interesting features. Section 4 presents our conclusions.

2 Main Result

2.1 The Backward Payoff-Keeping Behavior

Although our purpose is to define plausible equilibria, we need some preparation. The maximin criteria are usually defined in a strategic form game, and thus we need to modify them for an extensive form game. Hence, we initially define the criteria of actions in the extensive form game corresponding to the maximin criteria, named the backward payoff-keeping behavior.[5]

Definition 2.1 Consider a finite game.[6] The **kept value** for player i at a history h, denoted by $x_i(h)$, is defined recursively. (The function $P(h)$ denotes the player function and c denotes the nature.)

If h is terminal, then $x_i(h)$ is simply the payoff of player i at h.

Next, suppose that h is not terminal and $P(h) \neq i$. Let $V(h)$ be the set of all histories $h' = (h, a)$, where $a \in A(h)$, and suppose that for any $h' \in V(h)$, the kept value $x_i(h')$ has already been defined.

If $P(h) = c$, then $x_i(h)$ is the expectation of $x_i(h')$ over $V(h)$.

If $P(h) = j$ and $i \neq j \neq c$, then

$$x_i(h) = \min_{h' \in V(h)} x_i(h').$$

[5]Although there are many definitions of the extensive game, we basically follow Ch.11 of Osborne and Rubinstein [6].

[6]In this paper, the terminology "finite game" means a game that has the following two properties: (1) the length of the history $h \in H$ is bounded by some finite number, and (2) for every history $h \in H$, the set $A(h)$ of possible choices is finite.

Finally, suppose that $P(h) = i$ and I is the information set with $h \in I$. Suppose also that for any $h' \in I$ and any $a \in A(h)$, the kept value $x_i((h', a))$ has already been defined.[7] Then,

$$x_i(h) = \max_{a \in A(h)} \min_{h' \in I} x_i((h', a)),$$

and let $V^b(h)$ be the set of all actions $a \in A(h)$ such that the following equality holds:

$$\min_{h' \in I} x_i((h', a)) = x_i(h).$$

We call an element of $V^b(h)$ the **backward payoff-keeping behavior** at h.

Note that, the above definitions are not well-defined if several assumptions of the game do not hold. Clearly, in any game with an infinite time horizon, we cannot define the kept value. Additionally, in the class of imperfect recall games, there are many games in which the kept value cannot be defined at several nodes. Figure 1 shows such an example.

In Fig. 1, there are two information sets such that each one is an ancestor of another one. However, there exists a game in which kept value cannot be defined even if the above phenomenon is prohibited. For example, suppose that $N = \{1, 2\}$, and at time 1, player 1 chooses A, B, or C. At times 2 and 3, player 2 chooses N or S, and the information partition of player 2 is

$$\{\{A, BN, BS\}, \{B, CN, CS\}, \{C, AN, AS\}\}.$$

Fig. 1 A game in which the kept value cannot be defined

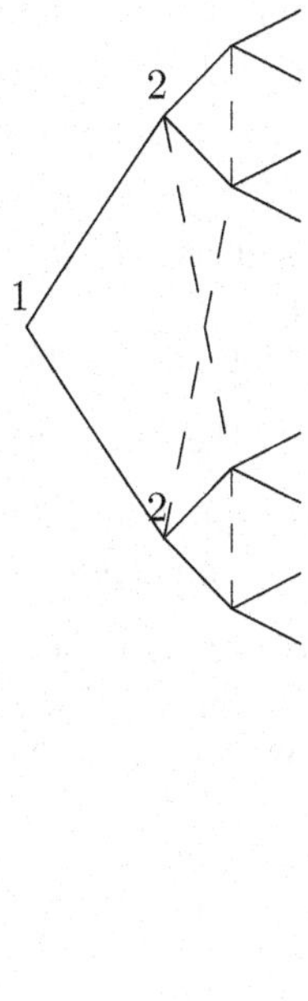

[7]Note that, because $h, h' \in I$, we must have $A(h) = A(h')$.

Then, the kept value of player 2 cannot be defined in this case. Actually, to define $x_2(A)$, the value $x_2(AN)$ is needed. Because AN is included in the same information set as C, to define $x_2(AN)$, the value $x_2(CN)$ is needed. Again, because CN is included in the same information set as B, to define $x_2(CN)$, the value $x_2(BN)$ is needed, which is the same as $x_2(A)$. Therefore, we have that to define $x_2(A)$, the value $x_2(A)$ itself is needed, which is impossible.

To avoid such a case, we should define a notion to ensure the well-definedness of the kept value. For any two information sets I, J of player i, we write $I \succ_i J$ if there exists $h \in J$ and $(h, h') \in I$.[8] We say that a game has a **no-cycle information structure** if $\succ_i$ is acyclic for every i.

Clearly, every perfect information game has a no-cycle information structure. Actually, the following proposition holds.

Proposition 2.1 *Every perfect recall game has a no-cycle information structure.*

Proof First, remember the definition of the perfect recall game. Let $h = (a_1, \ldots, a_\ell)$ be a history and for $k = 0, \ldots, \ell$, $(a_1, \ldots, a_k) \in I_k$, where I_k is an information set. We assume that $P(h) = i$. Let $\{k_1, \ldots, k_p\}$ denotes the set of all k such that I_k is an information set of i. We define $X_i(h) = \{I_{k_1}, a_{k_1+1}, \ldots, I_{k_p-1}, a_{k_p-1}, I_{k_p}\}$. This function X_i is called the record function for player i, and the game is perfect recall if $h, h' \in I$ and $P(h) = i$, then $X_i(h) = X_i(h')$.

Suppose the game does not have a no-cycle information structure. Then, there exists a finite sequence of information sets $I_1, \ldots, I_k$ of player i such that $h_j \in I_j$ and

$$I_1 \succ_i I_2 \succ_i \ldots \succ_i I_k \succ_i I_1.$$

By definition, there exist $h_j^1, h_j^2 \in I_j$ and h_j^3 such that $h_j^1 = (h_{j+1}^2, h_j^3)$ for every $j = 1, \ldots, k$, where h_{k+1}^2 means h_1^2. If the game is perfect recall, the value of the record function $X_i(h_j^1)$ coincides with $X_i(h_j^2)$. Therefore, we have that there exist $h_3^* \in I_3$ and h_2^+ such that $h_2^2 = (h_3^*, h_2^+)$, and consequently, there exists a finite sequence $h_1^*, h_2^*, \ldots, h_k^*, h_{k+1}^*$ and $h_1^+, \ldots, h_k^+$ such that $h_j^* \in I_j$, $h_j^* = (h_{j+1}^*, h_j^+)$ for $j = 1, \ldots, k$, and $h_{k+1}^* \in I_1$. This implies that $X_i(h_1^*) \neq X_i(h_{k+1}^*)$, a contradiction. ■

The next proposition ensures that our definition of the kept value is meaningful.

Proposition 2.2 *The kept value $x_i(h)$ is well-defined at any history $h \in H$ if and only if the game has a no-cycle information structure.*

Proof Suppose that the game has a no-cycle information structure. First, choose any history h with $P(h) = i$, and let I be the information set of i that includes h. We will show that the kept value $x_i(h)$ can be defined if $x_i((h', h''))$ can be defined

[8] If $h = (a_1, \ldots, a_k)$ and $h' = (a_{k+1}, \ldots, a_\ell)$, then (h, h') is an abbreviation of $(a_1, \ldots, a_\ell)$.

for any history (h', h'') such that $h' \in I$ and $P((h', h'')) = i$. It suffices to show that $x_i((h', h''))$ can be defined for every history (h', h'') such that $h' \in I$ and the length of h'' is greater than 0. We use mathematical induction on k with the length $n-k$ of h'' such that $(h', h'') \in H$ and $h' \in I$, where n is the maximal length of such h''.[9] If $k = 0$, then (h', h'') is terminal, and thus $x_i((h', h''))$ can be defined. Suppose that $x_i((h', h''))$ can be defined for any $(h', h'') \in H$ such that $h' \in I$ and the length of h'' is greater than $n-k$. Choose any h'' such that $(h', h'') \in H$, $h' \in I$ and the length of h'' is equal to $n-k$. If (h', h'') is terminal, then $x_i((h', h''))$ can be defined. If $P((h', h'')) = i$, then $x_i((h', h''))$ can be defined by our initial assumption. Otherwise, either $P((h', h'')) = c$ or $i \neq P((h', h'')) \neq c$. If the former holds, then $x_i((h', h''))$ is the expectation of $x_i((h', h'', a))$ over $A((h', h''))$. If the latter holds, then $x_i((h', h''))$ is the minimum of $x_i((h', h'', a))$ with respect to $a \in A((h', h''))$. In both cases, because the length of (h'', a) is $n-k+1$, $x_i((h', h'', a))$ is defined by the assumption of the mathematical induction, and thus $x_i((h', h''))$ can be defined. Hence, our claim holds.

Now, suppose that $x_i(h)$ cannot be defined for some $h \in H$. Let $\mathscr{I}^*$ be the set of all information set of player i such that $x_i(h)$ is undefined for $h \in I$. Because $\mathscr{I}^*$ is finite and $\succ_i$ is acyclic, there exists a maximal element $I^* \in \mathscr{I}^*$ with respect to $\succ_i$. If $h \in I^*$ and $(h, h') \in I$ for some information set I of player i, then $I \succ_i I^*$ and thus $I \notin \mathscr{I}^*$. Therefore, the kept value $x_i((h, h'))$ can be defined. As we argued above, we can define $x_i(h)$ for $h \in I^*$, a contradiction. Therefore, if the game has a no-cycle information structure, then the kept value is well-defined.

Next, suppose that $\succ_i$ has a cycle: that is,

$$I_1 \succ_i I_2 \succ_i \ldots \succ_i I_m \succ_i I_1.$$

Choose any $h_j \in I_j$. Then, to define $x_i(h_j)$, we need the value $x_i(h_{j-1})$, where $j-1 = m$ if $j = 1$. Thus, to define $x_i(h_1)$, the value $x_i(h_1)$ itself is needed, which is impossible. This completes the proof. ■

In the rest of this paper, we assume unless otherwise stated that the game is finite and has a no-cycle information structure.

2.2 The Plausible Equilibria

We can now define our refinement of NE.

Consider an extensive form game Γ. Let $u_i(s)$ denote the payoff of player i when the strategy profile $s = (s_1, \ldots, s_n)$ is realized. Suppose that s^* is an NE of Γ. Define $T^{s^*}_{i,k}$ as the set of all strategy profiles $t = (t_1, \ldots, t_n)$ that satisfies the following property. Let J be an information set owned by $j \neq i$. For each

[9] Note that $n > 0$ because h is not terminal.

$h = (a_1, \ldots, a_\alpha) \in J$, let $q(h)$ be the number of $\beta \leq \alpha$ such that if J' is the information set including $(a_1, \ldots, a_{\beta-1})$ owned by j' with $j \neq j' \neq c$, then a_β is different from the choice of $s^*_{j'}$. In other words, $q(h)$ is the number of past actions of other players than j that is inconsistent with the given strategy profile s^*. Define $p(J) = \min_{h \in J} q(h)$. If $p(J) \leq k$, then t_j must choose the same action as that chosen by s^*_j. If $p(J) > k$, then t_j must choose some backward payoff-keeping behavior at J. t_i is an arbitrary strategy of i.

Definition 2.2 The **degree of implausibility**(DIP) of an NE s^* is K if and only if K is the maximal number of k such that $u_i(s^*) \geq u_i(t)$ for any $i \in N$ and any $t \in T^{s^*}_{i,k}$. An NE s^* is **plausible** if the DIP of this NE is zero.

We interpret the chain store paradox as the possible existence of doubtful SPEs. The notion of plausible equilibria solves this problem: if an NE is plausible, then this NE is credible, and thus we can avoid the chain store paradox.

Note that the unique SPE of Selten's chain store game is not plausible. Here, we confirm this result. Let s^* be the unique SPE of this game. Then player 0 chooses C at every node, and the other players choose E at every node. Now, let t_i be the following:

(i) t_0 is "always A."
(ii) $t_1, \ldots, t_{11}$ is "always E," and for $i > 11$, t_i is "choose E if A or N is chosen less than twelve times, and choose N otherwise."

Then $t = (t_0, \ldots, t_{20}) \in T^{s^*}_{0,11}$. We can easily compute

$$u_0(t) = 45 > 40 = u_0(s^*),$$

and thus the DIP of s^* is greater than 11. In fact, we can easily show that the DIP of this s^* is exactly 12.

In contrast, there is an NE s^+ of Selten's game for which the DIP is 1. In this NE, s^+_0 is "always A," and s^+_i is "always N." Note that in this NE, player 0 gains the highest payoff in this game. Therefore, $u_0(s^+) \geq u_0(t)$ for all $t \in T^{s^+}_{0,1}$. Next, for player $i > 0$, either $u_i(t) = 1$ or $u_i(t) = 0$ whenever $t \in T^{s^+}_{i,1}$, and $u_i(s^+) = 1$. Meanwhile, $u_i(t) = 2$ for $t \in T^{s^+}_{i,0}$, where t_i is "always E." Therefore, the DIP of s^+ is 1.

Note that s^+ is very similar to the "deterrence theory" in Selten's explanation. Thus, we consider the notion of DIP to resolve Selten's chain store paradox. In Selten's game, the DIP of the unique SPE is too higher than an NE, and thus it is implausible.

In Selten's game, there is no plausible equilibrium. Therefore, plausible equilibria may be absent even if the game has perfect information. If there is no plausible equilibrium, then an NE with a least DIP may be substituted for the plausible equilibrium. In Selten's game, the above s^+ satisfies this requirement.

By definition, plausible equilibria tend to be absent when $T^{s^*}_{i,0}$ grows. Since $T^{s^*}_{i,0}$ is larger when the information partition is finer, games with perfect information seem to rarely ensure the existence of plausible equilibria. Additionally, plausible equilibria tend to be absent when the interests of players conflict. In fact, the following proposition holds.

Proposition 2.3 *For an NE s^*, if $u_i(s^*) \geq u_i(s)$ for any i and any strategy profile s, then s^* is plausible.*

Proof By assumption,

$$u_i(s^*) \geq u_i(t)$$

for every $t \in T_{i,0}^{s^*}$, and thus our claim is correct. ■

Therefore, one might consider that plausible equilibria to usually be absent for games with perfect information or when the players' interests conflict deeply.

However, the following theorem completely contradicts the above intuition.

Theorem 2.1 *Let Γ be any two-person finite perfect information zero-sum game,*[10] *and let s^* be any SPE of Γ. Then, s^* is plausible.*

Proof First, we introduce a lemma that helps to prove the theorem.

Lemma 2.1 *Let s_i^b be a strategy of player i that consists solely of backward payoff-keeping behaviors. Then s_i^b is a maximin strategy of player i. Moreover, the kept value of player i at any h is the maximin value of player i in the subgame whose root is h.*

Proof We will prove this lemma by mathematical induction with respect to the maximum length of histories n.

Suppose $n = 0$. Then, the set H of all histories is $\{\emptyset\}$, and every player has the strategy set $\{\emptyset\}$. Thus, $s_i^b = \emptyset$ and it is clearly the maximin strategy. Moreover, the kept value of the root is simply the payoff of player i, which is clearly the maximin value.

Next, assume that the lemma is correct when $n \leq m - 1$ and that Γ is any game where $n = m$. For history $h \neq \emptyset$, the induction hypothesis implies that $x_i(h)$ is the maximin value of the subgame whose root is h. Therefore, it suffices to show that s_i^b is a maximin strategy of player i and $x_i(\emptyset)$ is the maximin value of player i.

Now, let $A(\emptyset) = \{a_1, \ldots, a_N\}$, Γ^k be the subgame whose root is a_k, s_i^k (resp. $s_i^{b,k}$) be the restriction of s_i (resp. s_i^b) to Γ^k, and u_i^k be the payoff function of player i at Γ^k. By the induction hypothesis, $s_i^{b,k}$ is the maximin strategy of Γ^k and $x_i(a_k)$ is the maximin value of Γ^k. Therefore,

$$x_i(a_k) = \max_{s_i^k} \min_{s_j^k} u_i\left(s_i^k, s_j^k\right) = \min_{s_j^k} u_i\left(s_i^{b,k}, s_j^k\right).$$

[10] That is, the sum of the payoff at any terminal node is always zero.

First, suppose that $P(\emptyset) = i$. Then, s_i^b chooses an element of the argmax of $x_i(a_k)$ at $\emptyset$, which is denoted by a_{k^*}. Then,

$$\begin{aligned}\min_{s_j} u_i\left(s_i^b, s_j\right) &= \min_{s_j^{k^*}} u_i\left(s_i^{b,k^*}, s_j^{k^*}\right)\\ &= x_i(a_{k^*})\\ &= \max_k x_i(a_k)\\ &= \max_k \max_{s_i^k} \min_{s_j^k} u_i\left(s_i^k, s_j^k\right)\\ &= \max_{s_i} \min_{s_j} u_i(s_i, s_j),\end{aligned}$$

and thus, s_i^b is the maximin strategy of Γ. Moreover,

$$x_i(\emptyset) = \max_k x_i(a_k) = \max_{s_i} \min_{s_j} u_i(s_i, s_j).$$

Therefore, the lemma is correct in this case.

Second, suppose that $P(\emptyset) = j$, where $i \neq j \neq c$. Then, for any strategy s_i of player i,

$$\begin{aligned}\min_{s_j} u_i\left(s_i^b, s_j\right) &= \min_k \min_{s_j^k} u_i\left(s_i^{b,k}, s_j^k\right)\\ &\geq \min_k \min_{s_j^k} u_i\left(s_i^k, s_j^k\right)\\ &= \min_{s_j} u_i(s_i, s_j),\end{aligned}$$

and thus s_i^b is the maximin strategy of Γ. Moreover,

$$x_i(\emptyset) = \min_k x_i(a_k) = \min_k \min_{s_j^k} u_i\left(s_i^{b,k}, s_j^k\right) = \min_{s_j} u_i\left(s_i^b, s_j\right).$$

Hence, $x_i(\emptyset)$ is the maximin value, as claimed.

Finally, suppose that $P(\emptyset) = c$ and p_k denotes the probability of choosing a_k. Then, for any strategy s_i of player i,

$$\begin{aligned}\min_{s_j} u_i\left(s_i^b, s_j\right) &= \sum_k p_k \min_{s_j^k} u_i\left(s_i^{b,k}, s_j^k\right)\\ &\geq \sum_k p_k \min_{s_j^k} u_i\left(s_i^k, s_j^k\right)\\ &= \min_{s_j} u_i(s_i, s_j),\end{aligned}$$

and thus, s_i^b is also the maximin strategy. Moreover,

$$\begin{aligned} x_i(\emptyset) &= \sum_k p_k x_i(v_k) \\ &= \sum_k p_k \min_{s_j^k} u_i^k\left(s_i^{b,k}, s_j^k\right) \\ &= \min_{s_j} u_i\left(s_i^b, s_j\right) \\ &= \max_{s_i} \min_{s_j} u_i(s_i, s_j), \end{aligned}$$

which completes the proof of the lemma. ■

We now prove the theorem. Choose any SPE s^* of the game and take any $t = (t_i, t_j) \in T_{i,0}^{s^*}$. For any history $h \in H$, let Γ^h denote the subgame starting from h, t_k^h (resp. $s_k^{*,h}$) be the restriction of t_k (resp. s_k^*) to Γ^h, and u_k^h be the payoff function of Γ^h, where $k = i$ or $k = j$.

First, choose any $h \in H$ and suppose that the realized play by t is indistinguishable from that by s^* for player j on Γ^h. Then the choice of t_j^h at any realized history is identical to that of $s_j^{*,h}$. Hence,

$$u_i^h\left(t_i^h, t_j^h\right) = u_i^h\left(t_i^h, s_j^{*,h}\right) \leq u_i^h\left(s_i^{*,h}, s_j^{*,h}\right),$$

since $s_i^{*,h}$ is a best response of $s_j^{*,h}$.

Second, suppose that $u_i(t) > u_i(s^*)$. There exists a history h that is realized by both s^* and t with positive probability and such that $u_i^h(t^h) > u_i^h(s^{*,h})$.[11] Let V be the set of such histories and h^* be a member of V with maximal length. Note that h^* is not terminal. By definition of h^*, there exists an action $a \in A(h^*)$ such that for $w = (h^*, a)$, $u_i^w(t^w) > u_i^w(s^{*,w})$ and a is realized by t with positive probability. Then, $w \notin V$ by definition of h^*, and thus a is not realized by s^*. Because $t \in T_{i,0}^{s^*}$, this implies that $P(h^*) = i$. By the previous lemma, the strategy t_j^w of player j in Γ^w is the maximin strategy, and thus satisfies

$$\min_{s_i^w} u_j^w(s_i^w, t_j^w) = \max_{s_j^w} \min_{s_i^w} u_j^w(s_i^w, s_j^w).$$

Since the game is zero-sum,

$$-\min_{s_i^w} u_j^w(s_i^w, t_j^w) = \max_{s_i^w} u_i^w(s_i^w, t_j^w)$$

[11] For example, $\emptyset$ satisfies this condition.

and

$$-\max_{s_j^w}\min_{s_i^w} u_j^w(s_i^w, s_j^w) = \min_{s_j^w}\max_{s_i^w} u_i(s_i^w, s_j^w).$$

Therefore, we have

$$\max_{s_i^w} u_i^w(s_i^w, t_j^w) = \min_{s_j^w}\max_{s_i^w} u_i^w(s_i^w, s_j^w).$$

Clearly,

$$\max_{s_i^w} u_i^w(s_i^w, t_j^w) \geq u_i^w(t_i^w, t_j^w).$$

Since s^* is SPE and the game is zero-sum two-person, we also have

$$\min_{s_j^w}\max_{s_i^w} u_i^w\left(s_i^w, s_j^w\right) = u_i^w\left(s_i^{*,w}, s_j^{*,w}\right).$$

Hence,

$$u_i^w\left(s_i^{*,w}, s_j^{*,w}\right) \geq u_i^w\left(t_i^w, t_j^w\right),$$

which contradicts the definition of w. This completes the proof. ■

3 Applications

3.1 The Coordination Game

Consider the 2-repeated game with the following stage game.

	C	D
A	(3,3)	(0,0)
B	(0,0)	(1,1)

Then, both $((A,\ \text{always}\ A), (C,\ \text{always}\ C))$ and $((B,\ \text{always}\ B), (D,\ \text{always}\ D))$ are SPE. Proposition 2.3 says that the former is plausible. Meanwhile, it can easily be shown that the latter is not plausible.

More generally, consider the following game where $a, b > 0$.

	C	D
A	(a,a)	(0,0)
B	(0,0)	(b,b)

Then both (A, C) and (B, D) are NE of this game. Now, we say an NE (X, Y) is **steady-state plausible** if, for any N, (always take X, always take Y) is plausible in an N-repeated game with this stage game. Then it can easily be verified that (A, C) is steady-state plausible if and only if $a \geq b$.

Hence, the notion of steady-state plausibility removes the bad equilibrium. The interpretation of this feature is the following. Suppose that a bad equilibrium tends to become the steady state. If player 1 despises this equilibrium, then he/she can change his/her choice. If he/she changes his/her choice, the payoff of this stage is 0. However, in the following stage, player 2 can adapt his/her choice to player 1. Thus, the payoff increases. If N is sufficiently large, then their long-run payoffs can be improved. Therefore, this equilibrium is not plausible.

In contrast, the good equilibrium remains steady-state plausible, since no change of a player's choice improves his/her payoffs.

3.2 *The Battle of Sexes*

Consider the following game:

	C	D
A	(2,1)	(0,0)
B	(0,0)	(1,2)

Then it can easily be verified neither (A, C) nor (B, D) are steady-state plausible. This feature indicates that in this situation, conflict tends to occur and no equilibrium seems to be the steady state.

3.3 *Cournot Oligopoly*

Consider a simple linear Cournot game. Suppose that the inverse demand is $p(y) = 1500 - y$ and the cost of i is $c_i(y_i) = 300y_i$. The unique NE of this game is $(400, 400)$. Next, consider 2-repeated game with this stage game. Define

$$s^* = ((400, \text{ always } 400), (400, \text{ always } 400)),$$

$$t = ((401, \text{ always } 600), (400, \text{ always } 0 \text{ except player 1 takes 400 in stage 1})).$$

Then we can verify that $t \in T_{1,0}^{s^*}$ and $u_1(t) > u_1(s^*)$. Hence s^* is not plausible.

This seems to be strange, since t is a very strange strategy profile. In first stage, player 1 changes his choice very small. Then player 2 follows the maximin behavior in second stage, which is 0. This does not seem to be persuasive.

This example shows that if the strategy space is very wide, then the notion of plausible equilibria might be too strict. Even if an SPE is not plausible, the reason why it is not plausible might be absurd.

We should mention that this argument does not reduce the conviction of plausible equilibria. This argument only tells us that some non-plausible SPEs might be credible enough.

4 Conclusion

In this paper, we have proposed a refinement of SPE in which the phenomenon similar to the chain store paradox does not occur. Further, we have verified that two-person zero-sum games must have this refinement.

Future tasks include finding a broader class of games than zero-sum games in which this refinement is not absent and to construct a computation method for this refinement.

Acknowledgements We are grateful to Toshiyuki Komiya, Mikio Nakayama, Eishiro Takeda, Masayuki Yao, and an anonymous reviewer for their worthwhile comments and suggestions.

References

1. Fudenberg D, Tirole J (1991) Game theory. MIT Press, Cambridge, MA
2. Kreps DM, Wilson R (1982) Reputation and imperfect information. J Econ Theory 27:253–279
3. Kreps DM, Wilson R (1982) Sequential equilibria. Econometrica 50:863–894
4. Kreps DM, Milgrom P, Roberts J, Wilson R (1982) Rational cooperation in the finitely repeated prisoners' dilemma. J Econ Theory 27:245–252
5. Milgrom P, Roberts J (1982) Predation, reputation, and entry deterrence. J Econ Theory 27:280–312
6. Osborne MJ, Rubinstein A (1994) A course in game theory. MIT Press, Cambridge, MA
7. Scherer F (1980) Industrial market structure and economic performance, 2nd edn. Rand McNally College Publishing Company, Chicago
8. Selten R (1978) The chain store paradox. Theor Decis 9:127–159

A Unified Approach to Convergence Theorems of Nonlinear Integrals

Jun Kawabe

Abstract There are several types of nonlinear integrals with respect to nonadditive measures, such as the Choquet, Šipoš, Sugeno, and Shilkret integrals. In order to put those integrals into practical use and aim for application to various fields, it is indispensable to establish convergence theorems of such nonlinear integrals. However, they have individually been discussed for each of the integrals up to the present. In this article, several important convergence theorems of nonlinear integrals, such as the monotone convergence theorem, the bounded convergence theorem, and the Vitali convergence theorem, are formulated in a unified way regardless of the types of integrals.

Keywords Nonadditive measure · Nonlinear integral · Integral functional · Convergence theorem · Perturbation

Article type: Research Article
Received: September 27, 2017
Revised: January 10, 2018

1 Introduction

Nonadditive measures are monotonically increasing set functions that are not necessarily additive, and nonlinear integrals are the integrals with respect to nonadditive

---

JEL Classification: C02
Mathematics Subject Classification (2010): Primary 28-02, 28E10; Secondary 28A10, 28A25

J. Kawabe (✉)
Division of Mathematics and Physics, Faculty of Engineering, Shinshu University, Nagano, Japan
e-mail: jkawabe@shinshu-u.ac.jp

S. Kusuoka, T. Maruyama (eds.), *Advances in Mathematical Economics*, Advances in Mathematical Economics 22, https://doi.org/10.1007/978-981-13-0605-1_4

measures. There had already been fragmentary study on nonadditive measures and nonlinear integrals, but its importance was reaffirmed by a series of experiments in 1961, known as the "Ellsberg Paradox" in decision theory in which people's choices violate the postulates of subjective expected utility [6]. After that, by using the Choquet and Sugeno integrals with respect to nonadditive measures, expected utility theory has been reformulated and applied to various problems in decision-making [9, 10] and economics of pessimism and optimism [27].

The starting point for a systematic study of nonadditive measures and nonlinear integrals is a work on fuzzy measures by Sugeno [36] in 1974 in terms of applications to engineering and a work on submeasures by Dobrakov [5] interestingly in the same year in terms of purely mathematical interest. After that, a wide range of studies, both theory and application, has been made by many mathematicians, engineers, and economists, and now it turns out that various important theorems in ordinary measure theory can be established under practical and weaker additivity and continuity of measures. Early in the 1990s, those studies were summarized in the textbooks by Wang and Klir [41, 42], Denneberg [4], and Pap [28]. In this way, the study of nonadditive measures and nonlinear integrals is based on steady demands from researchers in the field of engineering and social sciences and considered as a new theory in which "additivity" is removed from measure theory and "linearity" is removed from integration theory.

Nonlinear integrals are important in terms of expected utility theory, subjective evaluation problem, and the refinement of measure and integration theory, in which the Lebesgue integral may not be a reasonable integral due to the lack of the additivity of measures. Moreover, in mathematical theory, addition and multiplication are basic binary operations, but from the engineering aspect, in addition to those, lattice operations such as supremum and infimum are often used. Therefore, depending on each specific problem in application fields, an appropriate integral is selected from among the Choquet integral defined by addition and multiplication, the Sugeno integral defined by supremum and infimum, and the Shilkret integral defined by supremum and multiplication.

In order to put those nonlinear integrals into practical use and aim for application to various fields, it is indispensable to establish convergence theorems assuring that the limit of the integrals of a sequence of functions is the integral of the limit function. In fact, in the engineering field those convergence theorems are considered to imply the robustness, the stability, and the non-chaotic state of aggregation processes. However, they have individually been discussed for each of the nonlinear integrals up to the present, so that the formulations of theorems and their proof methods deeply depend on the definition and properties peculiar to each integral.

In this article, a unified approach to convergence theorems of nonlinear integrals is introduced based on a series of papers [12–15, 17–20]. In Sect. 2 notation and terminologies are prepared, and in Sect. 3 typical nonlinear integrals such as the Choquet, Šipoš, Sugeno, and Shilkret integrals are introduced. In Sect. 4 nonlinear integrals are considered as nonlinear integral functionals defined on an appropriate domain, and several properties common to those integrals are described in terms of the integral functionals. Among them, the perturbation of integral functionals plays an essential role in our unified approach to convergence theorems. In Sect. 5, by using the properties of integral functionals given in Sect. 4, in particular, the perturbation of integral functionals, some of the important convergence theorems of nonlinear integrals, such as the monotone convergence theorem, the bounded convergence theorem, and the Vitali convergence theorem, are formulated in a unified way regardless of the types of integrals. Section 6 contains concluding remarks on the significance of the study of the Šipoš integral and a future task.

2 Preliminaries

Let X be a nonempty set and $\mathscr{A}$ a field of subsets of X. Let $\mathbb{R} = (-\infty, \infty)$ and $\mathbb{N}$ denote the set of all real numbers and the set of all natural numbers, respectively. Let $\overline{\mathbb{R}} := [-\infty, \infty]$ be the set of all extended real numbers with usual total order and algebraic structure. It is explicitly assumed that

$$(\pm\infty) \cdot 0 = 0 \cdot (\pm\infty) = 0,$$

since this proves to be convenient in measure and integration theory. In order that every nonempty subset A of $\overline{\mathbb{R}}$ always has the supremum and the infimum in $\overline{\mathbb{R}}$, let $\sup A := \infty$ if A is not bounded from above in $\mathbb{R}$ and $\inf A := -\infty$ if A is not bounded from below in $\mathbb{R}$. In addition, let $\inf \emptyset := \infty$.

For any $a, b \in \overline{\mathbb{R}}$, let $a \vee b := \max\{a, b\}$ and $a \wedge b := \min\{a, b\}$. For any family $\{f_\alpha\}_{\alpha\in\Gamma}$ of functions $f_\alpha \colon X \to \overline{\mathbb{R}}$, the supremum function $\sup_{\alpha\in\Gamma} f_\alpha$ and the infimum function $\inf_{\alpha\in\Gamma} f_\alpha$ are defined pointwise by

$$(\sup_{\alpha\in\Gamma} f_\alpha)(x) := \sup_{\alpha\in\Gamma} f_\alpha(x) \quad \text{and} \quad (\inf_{\alpha\in\Gamma} f_\alpha)(x) := \inf_{\alpha\in\Gamma} f_\alpha(x)$$

for each $x \in X$. In particular, for any $f, g \colon X \to \overline{\mathbb{R}}$, their supremum and infimum functions are denoted by $f \vee g$ and $f \wedge g$, that is, for each $x \in X$,

$$(f \vee g)(x) := f(x) \vee g(x) \quad \text{and} \quad (f \wedge g)(x) := f(x) \wedge g(x).$$

A function $f \colon X \to \overline{\mathbb{R}}$ is called *$\mathscr{A}$-measurable* if $\{f \geq t\}, \{f > t\} \in \mathscr{A}$ for every $t \in \overline{\mathbb{R}}$ and the set of all such functions is denoted by $\mathscr{F}(X)$. Let $\mathscr{F}_0(X) := \{f \in \mathscr{F}(X) \colon f \text{ is finite valued}\}$, $\mathscr{F}^+(X) := \{f \in \mathscr{F}(X) \colon f \geq 0\}$, and $\mathscr{F}_0^+(X) := \{f \in \mathscr{F}_0(X) \colon f \geq 0\}$. If $f, g \colon X \to \overline{\mathbb{R}}$ are $\mathscr{A}$-measurable and $c \in \mathbb{R}$, then the functions $f^+ := f \vee 0$, $f^- := (-f) \vee 0$, $|f| := f \vee (-f)$, cf, $f + c$, $(f - c)^+$, $f \vee g$, and $f \wedge g$ are all $\mathscr{A}$-measurable, and

$$f = f \wedge c + (f - c)^+$$

holds. Every function taking only a finite number of real numbers is called *simple* and $\mathscr{S}(X)$ denotes the set of all $\mathscr{A}$-measurable, simple functions on X. Let $\mathscr{S}^+(X) := \{f \in \mathscr{S}(X) : f \geq 0\}$.

For a sequence $\{f_n\}_{n\in\mathbb{N}} \subset \mathscr{F}(X)$ and $f \in \mathscr{F}(X)$, the symbol $f_n \to f$ denotes pointwise convergence, that is, $f_n(x) \to f(x)$ for every $x \in X$. It is written as $f_n \uparrow f$ if $\{f_n\}_{n\in\mathbb{N}}$ is increasing and as $f_n \downarrow f$ if $\{f_n\}_{n\in\mathbb{N}}$ is decreasing. For a net $\{f_\alpha\}_{\alpha\in\Gamma}$ of functions, $f_\alpha \to f$, $f_\alpha \uparrow f$, and $f_\alpha \downarrow f$ are similarly defined. Even in the case where $\mathscr{A}$ is a field, for any $f \in \mathscr{F}^+(X)$, there is an increasing sequence $\{h_n\}_{n\in\mathbb{N}} \subset \mathscr{S}^+(X)$ of simple functions such that $h_n \uparrow f$. If f is bounded, then h_n uniformly converges to f.

In this article, the extended real numbers are handled. Therefore, in order to avoid ambiguity of expression, when a positive number c can take the positive infinity, it is clearly represented as $c \in (0, \infty]$. In other words, $c > 0$ always means $c \in (0, \infty)$. The same notation is also used in other cases.

The definition of a nonadditive measure is mathematically quite simple.

Definition 2.1 A set function $\mu : \mathscr{A} \to [0, \infty]$ is called a *nonadditive measure* on X if it satisfies

(i) $\mu(\emptyset) = 0$ (boundedness from below),
(ii) $\mu(A) \leq \mu(B)$ whenever $A, B \in \mathscr{A}$ and $A \subset B$ (monotonicity).

A nonadditive measure is also referred to as a *monotone measure*, a *fuzzy measure*, and a *capacity* in the literature.

Instead of the monotonicity of Definition 2.1, if condition

(iii) whenever $A, B \in \mathscr{A}$ and $A \cap B = \emptyset$,

$$\mu(A \cup B) = \mu(A) + \mu(B) \quad \text{(finite additivity)}$$

is assumed, then μ is called a *finitely additive measure* and if condition

(iv) whenever $A_n \in \mathscr{A}$ $(n = 1, 2, \ldots)$, $\bigcup_{n=1}^{\infty} A_n \in \mathscr{A}$, and $A_i \cap A_j = \emptyset$ $(i \neq j)$,

$$\mu\left(\bigcup_{n=1}^{\infty} A_n\right) = \sum_{n=1}^{\infty} \mu(A_n) \quad (\sigma\text{-additivity})$$

is assumed, then μ is called a σ*-additive measure*. Every finitely additive measure and every σ-additive measure are nonadditive measures since they are monotone.

In what follows, $\mathscr{M}(X)$ denotes the set of all nonadditive measures on X. If $\mu(X) < \infty$, then μ is called *finite* and $\mathscr{M}_b(X)$ denotes the set of all finite nonadditive measures on X. For $\mu \in \mathscr{M}_b(X)$, its *dual* $\bar{\mu} : \mathscr{A} \to [0, \infty)$ is defined by

$$\bar{\mu}(A) := \mu(X) - \mu(X \setminus A), \quad A \in \mathscr{A}.$$

It is obvious that $\bar{\bar{\mu}} = \mu$. If μ is finitely additive, then $\bar{\mu} = \mu$.

For a sequence $\{A_n\}_{n\in\mathbb{N}} \subset \mathscr{A}$ and $A \in \mathscr{A}$, the symbol $A_n \uparrow A$ denotes that $\{A_n\}_{n\in\mathbb{N}}$ is increasing and $A = \bigcup_{n=1}^{\infty} A_n$, and the symbol $A_n \downarrow A$ denotes that $\{A_n\}_{n\in\mathbb{N}}$ is decreasing and $A = \bigcap_{n=1}^{\infty} A_n$. Let χ_A be the characteristic function of a set A.

The continuity of nonadditive measures can be defined in the same way as ordinary measures. A nonadditive measure $\mu\colon \mathscr{A} \to [0,\infty]$ is called *continuous from above* if $\mu(A_n) \to \mu(A)$ whenever $\{A_n\}_{n\in\mathbb{N}} \subset \mathscr{A}$, $A \in \mathscr{A}$, and $A_n \downarrow A$ and called *conditionally continuous from above* if $\mu(A_n) \to \mu(A)$ whenever $A_n \downarrow A$ and $\mu(A_1) < \infty$. It is called *continuous from below* if $\mu(A_n) \to \mu(A)$ whenever $\{A_n\}_{n\in\mathbb{N}} \subset \mathscr{A}$, $A \in \mathscr{A}$ and $A_n \uparrow A$. Furthermore, μ is called *continuous* if it is continuous from above and below and called *conditionally continuous* if it is conditionally continuous from above and continuous from below. It is obvious that the continuity from above and the conditional continuity from above coincide if $\mu(X) < \infty$.

The σ-additivity implies not the continuity from above but the conditional continuity from above. For instance, the Lebesgue measure m on $\mathbb{R}$ is not continuous from above since $A_n := \bigcup_{k=n}^{\infty}[k,\infty) \downarrow \emptyset$, $m(A_n) = \infty$ for all $n \in \mathbb{N}$, but $m(\emptyset) = 0$. Therefore, the concept of continuity from above is believed to be too strong, so that there is a tendency to avoid its use in ordinary measure theory. However, in nonadditive measure theory, where the additivity of measures is not a prerequisite, the *unconditionally* continuous infinite nonadditive measure $\mu = \theta \circ p$ on $\mathbb{R}$ is simply obtained by distorting a probability measure p on $\mathbb{R}$ by the function $\theta\colon [0,1] \to [0,\infty]$ defined as

$$\theta(t) := \begin{cases} \tan\left(\dfrac{\pi t}{2}\right) & \text{if } t \in [0,1) \\ \infty & \text{if } t = 1, \end{cases}$$

where $\theta \circ p(A) := \theta(p(A))$ for every Borel measurable subset A of $\mathbb{R}$. Therefore, the concept of continuity as well as that of conditional continuity is often a subject of study in nonadditive measure theory.

In general, if μ is not additive, then $\mu(N) = 0$ is not always equivalent to $\mu(X \setminus N) = \mu(X)$. Therefore, the concept of "almost everywhere" is defined in two ways depending on whether $\mu(N) = 0$ or $\mu(X \setminus N) = \mu(X)$ is adopted as the definition of null sets. In this article, standard definitions in ordinary measure theory are adopted when defining the notion of null sets, almost everywhere, almost everywhere convergence, convergence in measure, and so on; see textbooks [2, 11, 29]. For terminology and basic properties of nonadditive measures and nonlinear integrals, see [4, 28, 41, 42].

3 Nonlinear Integrals

Various types of nonlinear integrals are proposed in connection with a variety of concrete applications. Among them, the following four types of nonlinear integrals are important and of wide application. They are determined by the *μ-decreasing distribution function* of f defined by

$$G_{\mu,f}(t) := \mu(\{f \geq t\}), \quad t \in \mathbb{R}$$

and special cases of nonlinear integrals called *distribution-based integrals*.

Definition 3.2 Let $(\mu, f) \in \mathcal{M}(X) \times \mathcal{F}^+(X)$.

(1) The *Choquet integral* [3, 31] is defined by

$$\mathrm{Ch}(\mu, f) := \int_0^\infty \mu(\{f \geq t\})dt,$$

where the right-hand side is the Lebesgue integral or the improper Riemann integral.

(2) The *Šipoš integral* [33] is defined by

$$\mathrm{Si}(\mu, f) := \lim_{P \in \Delta^+} \sum_{i=1}^n (a_i - a_{i-1})\mu(\{f \geq a_i\}),$$

where Δ^+ is the directed set of all partitions of $[0, \infty]$ of the form $P = \{a_1, a_2, \ldots, a_n\}$ $(0 = a_0 < a_1 < \cdots < a_n < \infty)$.

(3) The *Sugeno integral* [30, 36] is defined by

$$\mathrm{Su}(\mu, f) := \sup_{t \in [0,\infty]} \left[t \wedge \mu(\{f \geq t\})\right].$$

(4) The *Shilkret integral* [32, 44] is defined by

$$\mathrm{Sh}(\mu, f) := \sup_{t \in [0,\infty]} \left[t \cdot \mu(\{f \geq t\})\right].$$

Remark 3.1 The equality $\mathrm{Ch}(\mu, f) = \mathrm{Si}(\mu, f)$ holds for any $(\mu, f) \in \mathcal{M}(X) \times \mathcal{F}^+(X)$ and both integrals are equal to the abstract Lebesgue integral if μ is σ-additive [33, 34]. Therefore, there is no difference between the Choquet and Šipoš integrals. In fact, the Šipoš integral is nothing but the improper Riemann integral of the decreasing nonnegative function $\mu(\{f \geq t\})$ written via a more elementary definition. Nevertheless, it is meaningful to consider the Šipoš integral in addition to the Choquet integral. Indeed, by exploring the Šipoš integral, it is simultaneously possible to construct both theories of the Lebesgue and Choquet integrals since the Šipoš integral can be defined and studied without knowledge of the Lebesgue integral.

Besides the abovementioned four distribution-based integrals, there are other important nonlinear integrals such as the pan integral by Yang [43] and the concave integral by Lehrer and Teper [21, 22].

Definition 3.3 Let $(\mu, f) \in \mathscr{M}(X) \times \mathscr{F}^+(X)$.

(1) The *pan integral* [43] with respect to the usual sum $+$ and multiplication $\cdot$ is defined by

$$\mathrm{Pan}(\mu, f) := \sup\left\{\sum_{i=1}^{n} r_i\mu(A_i) : \sum_{i=1}^{n} r_i\chi_{A_i} \le f,\ n \in \mathbb{N},\ A_i \in \mathscr{A},\right.$$
$$\left. r_i \ge 0,\ A_i \cap A_j = \emptyset\ (i \ne j),\ \cup_{i=1}^{n} A_i = X\right\}$$

(2) The *concave integral* [22] is defined by

$$\mathrm{Cav}(\mu, f) := \sup\left\{\sum_{i=1}^{n} r_i\mu(A_i) : \sum_{i=1}^{n} r_i\chi_{A_i} \le f,\ n \in \mathbb{N},\ r_i \ge 0,\ A_i \in \mathscr{A}\right\}$$

Remark 3.2

(1) The pan integral is introduced in [43] and in fact defined by using the pair $(\oplus, \otimes)$ of pan-addition $\oplus$ and pan-multiplication $\otimes$ on $[0, \infty]$. It coincides with the Sugeno integral when $(\oplus, \otimes) = (\vee, \wedge)$ and with the Shilkret integral when $(\oplus, \otimes) = (\vee, \cdot)$ [42, 43].
(2) In general, $\mathrm{Ch}(\mu, f) \le \mathrm{Cav}(\mu, f)$ and the equality holds if and only if μ is supermodular [22], while $\mathrm{Pan}(\mu, f) \le \mathrm{Cav}(\mu, f)$ and the equality holds if μ is subadditive. Therefore, $\mathrm{Ch}(\mu, f) = \mathrm{Pan}(\mu, f) = \mathrm{Cav}(\mu, f)$ if μ is additive and they coincide with the abstract Lebesgue integral if μ is σ-additive. In addition, $\mathrm{Pan}(\mu, f) \ge \mathrm{Ch}(\mu, f)$ if μ is subadditive and $\mathrm{Pan}(\mu, f) \le \mathrm{Ch}(\mu, f)$ if μ is superadditive [42].

The pan integral is determined by the collection of all finite measurable decompositions of X and contains the Sugeno, Shilkret, and Lebesgue integrals. On the other hand, the concave integral is determined by the collection of all finite families of measurable subsets of X and has the concavity, which means that the inequality

$$\mathrm{Cav}(\mu, \lambda f + (1-\lambda)g) \ge \lambda\mathrm{Cav}(\mu, f) + (1-\lambda)\mathrm{Cav}(\mu, g)$$

holds for any $f, g \in \mathscr{F}^+(X)$ and $\lambda \in [0, 1]$. This concavity might be interpreted as uncertainty aversion in the context of decision under uncertainty.

The pan and concave integrals are special cases of nonlinear integrals called *decomposition-based integrals* [7], but their discussions are saved for another time due to limitations of space.

4 Nonlinear Integral Functionals

In this chapter, several properties common to the Choquet, Šipoš, Sugeno, and Shilkret integrals are described together with related concepts. The main content in this and all subsequent chapters is a summary of the author's work [12–15, 17–20], so the citation of the individual papers is not specified from now on. The abovementioned papers and their references are of help for the readers.

Let $I\colon \mathcal{M}(X)\times\mathcal{F}^+(X)\to[0,\infty]$ be an *integral functional*, that is, it satisfies

(i) $I(\mu,0)=0$ for every $\mu\in\mathcal{M}(X)$,
(ii) $I(\mu,f)\le I(\mu,g)$ whenever $\mu\in\mathcal{M}(X)$, $f,g\in\mathcal{F}^+(X)$, and $f\le g$.

Definition 4.4 Let $I\colon \mathcal{M}(X)\times\mathcal{F}^+(X)\to[0,\infty]$ be an integral functional.

(1) I is called *upper marginal continuous* if for every $(\mu,f)\in\mathcal{M}(X)\times\mathcal{F}^+(X)$,

$$I(\mu,f)=\sup_{r>0} I(\mu,f\wedge r).$$

(2) I is called *lower marginal continuous* if for every $(\mu,f)\in\mathcal{M}(X)\times\mathcal{F}^+(X)$,

$$I(\mu,f)=\sup_{r>0} I(\mu,(f-r)^+).$$

(3) I is called *measure-truncated* if for every $(\mu,f)\in\mathcal{M}(X)\times\mathcal{F}^+(X)$,

$$I(\mu,f)=\sup_{s>0} I(s\wedge\mu,f),$$

where $(s\wedge\mu)(A):=s\wedge\mu(A)$ for every $A\in\mathcal{A}$ and $s>0$.

(4) I is called *horizontally subadditive* if for every $(\mu,f)\in\mathcal{M}(X)\times\mathcal{F}^+(X)$ and every $c>0$,

$$I(\mu,f)\le I(\mu,f\wedge c)+I(\mu,(f-c)^+)$$

and called *horizontally additive* if the equality holds.

(5) I is called *inner regular* if for every $(\mu,f)\in\mathcal{M}(X)\times\mathcal{F}^+(X)$,

$$I(\mu,f)=\sup\left\{I(\mu,h)\colon h\in\mathcal{S}^+(X),\ h\le f\right\}.$$

By definition, the restriction of I onto the set $\mathscr{M}(X) \times \mathscr{F}_b^+(X)$ is always upper marginal continuous, while that of I onto the set $\mathscr{M}_b(X) \times \mathscr{F}^+(X)$ is always measure-truncated.

For the Choquet integral

$$\mathrm{Ch}(\mu, f) := \int_0^\infty \mu(\{f \geq t\})dt, \quad (\mu, f) \in \mathscr{M}(X) \times \mathscr{F}^+(X),$$

each of the above characteristics has the following form:

- Upper marginal continuity:
$$\int_0^\infty \mu(\{f \geq t\})dt = \sup_{r>0} \int_0^r \mu(\{f \geq t\})dt$$
- Lower marginal continuity:
$$\int_0^\infty \mu(\{f \geq t\})dt = \sup_{r>0} \int_r^\infty \mu(\{f \geq t\})dt$$
- Measure-truncation:
$$\int_0^\infty \mu(\{f \geq t\})dt = \sup_{s>0} \int_0^\infty s \wedge \mu(\{f \geq t\})dt$$
- Horizontal additivity:
$$\int_0^\infty \mu(\{f \geq t\})dt = \int_0^c \mu(\{f \geq t\})dt + \int_c^\infty \mu(\{f \geq t\})dt$$
- Inner regularity:
$$\int_0^\infty \mu(\{f \geq t\})dt = \sup\left\{\int_0^\infty \mu(\{h \geq t\})dt : h \in \mathscr{S}^+(X), h \leq f\right\}$$

Proposition 4.1 *The integral functionals* Ch, Si, Su, *and* Sh *are upper marginal continuous, lower marginal continuous, measure-truncated, and inner regular. In addition,* Ch *and* Si *are horizontally additive, while* Su *and* Sh *are horizontally subadditive.*

The concepts of pseudo-addition and pseudo-difference are useful for representing the integral values of simple functions in a unified way regardless of various types of nonlinear integrals [1, 37].

Definition 4.5 A binary operation $\oplus \colon [0, \infty]^2 \to [0, \infty]$ is called a *pseudo-addition* if for any a, b, a', b', a_0, and b_0 in $[0, \infty]$, the following five conditions are satisfied:

(A1) $a \oplus b = b \oplus a$ (commutative law)
(A2) $(a \oplus b) \oplus c = a \oplus (b \oplus c)$ (associative law)
(A3) $a \oplus b \leq a' \oplus b'$ whenever $a \leq a'$ and $b \leq b'$ (monotonicity)
(A4) $a \oplus 0 = 0 \oplus a = a$ (neutral element)
(A5) It is continuous on $[0, \infty]^2$, that is, $\lim_{(a,b)\to(a_0,b_0)} a \oplus b = a_0 \oplus b_0$ (continuity).

The binary operation $\ominus \colon [0, \infty]^2 \to [0, \infty]$ defined by

$$a \ominus b := \inf\{x \in [0, \infty] \colon b \oplus x \geq a\}$$

for every $a, b \in [0, \infty]$ is called the *pseudo-difference* determined by a pseudo-addition $\oplus$.

Example 4.1

(1) Let $g \colon [0, \infty] \to [0, \infty]$ be an increasing bijection. The binary operation $\oplus \colon [0, \infty]^2 \to [0, \infty]$ defined by

$$a \oplus b := g^{-1}(g(a) + g(b))$$

for every $a, b \in [0, \infty]$ is a pseudo-addition and its pseudo-difference is given by

$$a \ominus b := \begin{cases} g^{-1}(g(a) - g(b)) & \text{if } a > b, \\ 0 & \text{if } a \leq b. \end{cases}$$

In particular, $a \oplus b := a + b$ is a pseudo-addition and $a \ominus b = a - b$ if $a > b$.

(2) The binary operation $\oplus \colon [0, \infty]^2 \to [0, \infty]$ defined by

$$a \oplus b := a \vee b$$

for every $a, b \in [0, \infty]$ is a pseudo-addition and its pseudo-difference is given by

$$a \ominus b := \begin{cases} a & \text{if } a > b, \\ 0 & \text{if } a \leq b. \end{cases}$$

For any pseudo-addition $\oplus$, every simple function $h \in \mathscr{S}^+(X)$ with $h(X) \setminus \{0\} = \{r_1, r_2, \ldots, r_n\}$ has a unique *standard $\oplus$-step representation*

$$h = \bigoplus_{i=1}^{n} (r_i \ominus r_{i-1}) \chi_{\{h \geq r_i\}},$$

where $n \in \mathbb{N}$ and $0 = r_0 < r_1 < \cdots < r_n < \infty$. In particular, if $\oplus = +, \vee$, then h is expressed by

$$h = \sum_{i=1}^{n} (r_i - r_{i-1}) \chi_{\{h \geq r_i\}} = \bigvee_{i=1}^{n} r_i \chi_{\{h \geq r_i\}}.$$

The following concepts of nonlinear functionals are necessary to calculate the integral values of simple functions.

Definition 4.6 Let $I\colon \mathscr{M}(X) \times \mathscr{F}^+(X) \to [0, \infty]$ be an integral functional.

(1) I is called *generative* if there is a function $\theta\colon [0, \infty]^2 \to [0, \infty]$ such that

$$I(\mu, r\chi_A) = \theta(r, \mu(A))$$

for every $\mu \in \mathscr{M}(X)$, $r \in [0, \infty]$, and $A \in \mathscr{A}$. In this case, θ is called a *generator* of I.

(2) I is called *elementary* if it is generative with generator θ and there is a pseudo-addition $\oplus\colon [0, \infty]^2 \to [0, \infty]$ such that

$$I\left(\mu, \bigoplus_{i=1}^{n} (r_i \ominus r_{i-1})\, \chi_{A_i}\right) = \bigoplus_{i=1}^{n} \theta\left(r_i \ominus r_{i-1}, \mu(A_i)\right)$$

for every $\mu \in \mathscr{M}(X)$, $n \in \mathbb{N}$, $r_1, \ldots, r_n \in (0, \infty)$, and $A_1, \ldots, A_n \in \mathscr{A}$ with $0 = r_0 < r_1 < \cdots < r_n < \infty$ and $A_1 \supset \cdots \supset A_n$.

Proposition 4.2 *The integral functionals* Ch, Si, Su, *and* Sh *have the following properties:*

(1) Ch *and* Si *are generative and elementary with generator* $\theta(a, b) = a \cdot b$, *and their pseudo-addition is* $a \oplus b = a + b$.
(2) Su *is generative and elementary with generator* $\theta(a, b) = a \wedge b$, *and its pseudo-addition is* $a \oplus b = a \vee b$.
(3) Sh *is generative and elementary with generator* $\theta(a, b) = a \cdot b$, *and its pseudo-addition is* $a \oplus b = a \vee b$.

The regulators of generative and elementary integral functionals usually have the following reasonable properties.

Definition 4.7 Let $\theta\colon [0, \infty]^2 \to [0, \infty]$ be a function of two variables.

(1) θ is called *of finite type* if $\theta(a, b) < \infty$ whenever $a, b \in [0, \infty)$.
(2) θ is called *of continuous type* if it is continuous on the set $D := [0, \infty]^2 \setminus \{(0, \infty), (\infty, 0)\}$.
(3) θ is called *limit preserving* if for any sequence $\{b_n\}_{n\in\mathbb{N}} \subset [0, \infty]$ and $b \in [0, \infty]$, $b_n \to b$ whenever $\theta(r, b_n) \to \theta(r, b)$ for every $r \in (0, \infty)$.

Proposition 4.3 *The functions* $\theta(a, b) := a \cdot b,\ a \wedge b$ *are of finite and continuous type and limit preserving.*

The following order relation among the pairs (μ, f) of set functions μ and functions f is already known as stochastic dominance, that is, stochastic ordering of random variables.

Definition 4.8 Let $\mu, \nu\colon \mathscr{A} \to [0, \infty]$ be set functions and $f, g \in \mathscr{F}(X)$. The pair (μ, f) is called *dominated* by the pair (ν, g) and written as $(\mu, f) \prec (\nu, g)$ if

$$\mu(\{f \geq t\}) \leq \nu(\{g \geq t\})$$

for every $t \in \mathbb{R}$.

Remark 4.3 The dominance $(\mu, f) \prec (\mu, g)$ is also called the first-order stochastic dominance and widely used in economics and finance together with the second-order and the third-order stochastic dominance; see a survey in [24].

In what follows, denote by Φ the set of all functions $\varphi\colon [0, \infty) \to [0, \infty)$ satisfying

$$\varphi(0) = \lim_{t \to +0} \varphi(t) = 0.$$

A function belonging to Φ is called a *control function*.

The *μ-essential boundedness constant* $\|f\|_\mu$ of a function $f \in \mathscr{F}(X)$ is the infimum of the set of all $r \in (0, \infty)$ satisfying

$$\mu(\{f \geq r\}) = 0 \quad \text{and} \quad \mu(\{f \geq -r\}) = \mu(X).$$

If $\|f\|_\mu < \infty$, then f is called *μ-essentially bounded*. Every bounded $f \in \mathscr{F}(X)$ is μ-essentially bounded and

$$\|f\|_\mu \leq \|f\| := \sup_{x \in X} |f(x)|.$$

It is obvious that the μ-essential boundedness constant $\|f\|_\mu$ is equal to the ordinary μ-essential supremum if f is nonnegative or μ is additive.

The perturbation of integral functionals introduced below plays an essential role in our unified approach to convergence theorems of nonlinear integrals. In fact, it successfully controls the change in the functional value $I(\mu, f)$ when the integrand is slightly shifted from f to $f + \varepsilon$ and its μ-decreasing distribution function is also slightly shifted from $\mu(\{f \geq t\})$ to $\mu(\{f \geq t\}) + \delta$.

Definition 4.9 Let $I\colon \mathscr{M}(X) \times \mathscr{F}^+(X) \to [0, \infty]$ be an integral functional.

(1) I is called *strongly monotone* (for short, *s-monotone*) if

$$I(\mu, f) \leq I(\mu, g)$$

whenever $\mu \in \mathscr{M}(X)$, $f, g \in \mathscr{F}^+(X)$, and $(\mu, f) \prec (\mu, g)$.

(2) I is called *perturbative* if there are families $\{\varphi_p\}_{p>0} \subset \Phi$ and $\{\psi_q\}_{q>0} \subset \Phi$ of control functions satisfying the following perturbation (P): For any $\mu \in \mathscr{M}(X)$, $f, g \in \mathscr{F}^+(X)$, $\varepsilon \geq 0$, $\delta \geq 0$, $p > 0$, and $q > 0$, the inequality

$$I(\mu, f) \leq I(\mu, g) + \varphi_p(\delta) + \psi_q(\varepsilon)$$

holds whenever $\|f\|_\mu < p$, $\mu(X) < q$, and $(\mu, f) \prec (\mu + \delta, g + \varepsilon)$.

Proposition 4.4 *The integral functionals* Ch, Si, Su, *and* Sh *are s-monotone and perturbative. Their control functions can be chosen as*

$$\varphi_p(t) = pt,\ pt,\ p \wedge t,\ pt \quad \text{and} \quad \psi_q(t) = qt,\ qt,\ q \wedge t,\ qt,$$

respectively.

Let $I \colon \mathscr{M}(X) \times \mathscr{F}^+(X) \to [0, \infty]$ be an integral functional. For each $\mu \in \mathscr{M}(X)$, the functional $I_\mu \colon \mathscr{F}^+(X) \to [0, \infty]$ defined by

$$I_\mu(f) := I(\mu, f), \quad f \in \mathscr{F}^+(X),$$

satisfies

(i) $I_\mu(0) = 0$,
(ii) $I_\mu(f) \le I_\mu(g)$ whenever $f, g \in \mathscr{F}^+(X)$ and $f \le g$

and called the *μ-integral functional* determined by I. Given $\mu \in \mathscr{M}(X)$, the μ-integral functional I_μ is called *upper marginal continuous*, *lower marginal continuous*, *measure-truncated*, *horizontally subadditive*, *horizontally additive*, and *inner regular* if the corresponding properties in Definition 4.4 hold for the fixed μ. For instance, I_μ is called *upper marginal continuous* if

$$I_\mu(f) = \sup_{r>0} I(\mu, f \wedge r)$$

for every $f \in \mathscr{F}^+(X)$ and called *measure-truncated* if

$$I_\mu(f) = \sup_{s>0} I(s \wedge \mu, f)$$

for every $f \in \mathscr{F}^+(X)$. Similarly, given $\mu \in \mathscr{M}(X)$, I_μ is called *generative*, *elementary*, *s-monotone*, and *perturbative* if the corresponding properties in Definitions 4.6 and 4.9 hold for the fixed μ. For instance, I_μ is called *generative* if there is a function $\theta \colon [0, \infty]^2 \to [0, \infty]$, which may depend on μ, such that

$$I_\mu(r\chi_A) = \theta(r, \mu(A))$$

for every $r \in [0, \infty]$ and $A \in \mathscr{A}$, and called *perturbative* if there are families $\{\varphi_p\}_{p>0} \subset \Phi$ and $\{\psi_q\}_{q>0} \subset \Phi$, which may depend on μ, satisfying the following perturbation $(P)_\mu$: For any $f, g \in \mathscr{F}^+(X)$, $\varepsilon \ge 0$, $\delta \ge 0$, $p > 0$, and $q > 0$, the inequality

$$I_\mu(f) \le I_\mu(g) + \varphi_p(\delta) + \psi_q(\varepsilon)$$

holds whenever $\|f\|_\mu < p$, $\mu(X) < q$, and $(\mu, f) \prec (\mu + \delta, g + \varepsilon)$.

Proposition 4.5 Let $I : \mathscr{M}(X) \times \mathscr{F}^+(X) \to [0, \infty]$ be an integral functional and $\mu \in \mathscr{M}(X)$. Then, if I is upper marginal continuous, lower marginal continuous, measure-truncated, horizontally subadditive, horizontally additive, inner regular, and s-monotone, so is I_μ, respectively. Moreover, If I is generative, elementary, and perturbative, so is I_μ with respect to the same regulator θ, pseudo-addition $\oplus$, and families $\{\varphi_p\}_{p>0}$ and $\{\psi_q\}_{q>0}$ of control functions as those of I, respectively.

5 Some Convergence Theorems of Nonlinear Integrals

The convergence theorems of nonlinear integrals, such as the monotone convergence theorem and the bounded convergence theorem, have individually been discussed for each of the integrals up to the present. Therefore, formulations of theorems and their proof methods deeply depend on the definition and properties peculiar to each integral. In this section, a unified approach to convergence theorems of nonlinear integrals is introduced from a functional analytic view by formulating those convergence theorems for integral functionals satisfying some of the properties in Sect. 4. Before going into details, we summarize below in tabular form the monotone increasing and decreasing convergence theorems and the Vitali convergence theorem, which are already known for each of the integrals.

In the rest of the article, let $(X, \mathscr{A})$ be a measurable space, $I : \mathscr{M}(X) \times \mathscr{F}^+(X) \to [0, \infty]$ an integral functional, and $\mu \in \mathscr{M}(X)$. The symbol Le denotes the integral functional defined by the abstract Lebesgue integral

$$\mathrm{Le}(\mu, f) := \int_X f \, d\mu$$

of a function $f \in \mathscr{F}^+(X)$ with respect to a σ-additive measure μ on $(X, \mathscr{A})$.
(I) The monotone increasing convergence theorem: Let $\{f_n\}_{n\in\mathbb{N}} \subset \mathscr{F}^+(X)$, $f \in \mathscr{F}^+(X)$, and $f_n \uparrow f$. Then $I_\mu(f_n) \to I_\mu(f)$ if f_n and μ satisfy the conditions in the table below. Here, "continuous ↑" is short for "continuous from below."

I	f_n	μ	References
Le	No condition	σ-additive	Levi [23]
Ch	No condition	Continuous ↑	Song and Li [35], Wang [40]
Si	No condition	Continuous ↑	Šipoš [33]
Su	No condition	Continuous ↑	Ralescu and Adams [30], Wang [39]
Sh	No condition	Continuous ↑	Zhao [44]

(II) The monotone decreasing convergence theorem: Let $\{f_n\}_{n\in\mathbb{N}} \subset \mathscr{F}^+(X)$, $f \in \mathscr{F}^+(X)$, and $f_n \downarrow f$. Then $I_\mu(f_n) \to I_\mu(f)$ if f_n and μ satisfy the conditions in the table below. Here, "cond. continuous ↓" is short for "conditionally continuous from above."

I	f_n	μ	References
Le	$\mathrm{Le}(\mu, f_1) < \infty$	σ-additive	Levi [23]
Ch	$\mathrm{Ch}(\mu, f_1) < \infty$	Cond. continuous $\downarrow$	Wang [40]
Si	$\mathrm{Si}(\mu, f_1) < \infty$	Cond. continuous $\downarrow$	Šipoš [33]
Su	$\mu(\{f_1 > \mathrm{Su}(\mu, f)\}) < \infty$	Cond. continuous $\downarrow$	Wang [39]
Sh	$\mu(\{f_1 > 0\}) < \infty$ and f_1 is μ-essentially bounded	Cond. continuous $\downarrow$	Zhao [45] Kawabe [17]

Recall that a sequence $\{f_n\}_{n\in\mathbb{N}} \subset \mathscr{F}_0(X)$ converges to $f \in \mathscr{F}_0(X)$ in μ-measure, which is written as $f_n \xrightarrow{\mu} f$, if

$$\lim_{n\to\infty} \mu(\{|f_n - f| > \varepsilon\}) = 0$$

for every $\varepsilon > 0$.
(III) The Vitali convergence theorem: Let $\{f_n\}_{n\in\mathbb{N}} \subset \mathscr{F}_0^+(X)$, $f \in \mathscr{F}_0^+(X)$, and $f_n \xrightarrow{\mu} f$. Then $I_\mu(f_n) \to I_\mu(f)$ if f_n and μ satisfy the conditions in the table below. Here, the autocontinuity of nonadditive measures is defined in Definition 5.11 below. Moreover, "unif. integrable for I_μ" is short for "uniformly integrable for I_μ" and defined in Definition 5.12 below.

I	f_n	μ	References
Le	Unif. integrable for Le_μ	Finite and σ-additive	Vitali [38]
Ch	Unif. integrable for Ch_μ	Finite and autocontinuous	Kawabe [19]
Si	Unif. integrable for Su_μ	Finite and autocontinuous	Kawabe [20]
Su	No condition	Autocontinuous	Wang [39]
Sh	Unif. integrable for Sh_μ	Finite and autocontinuous	Kawabe [20]

Remark 5.4 The conditions imposed on f_n and μ in tables (I) to (III) above cannot be removed.

As can be predicted by looking at the results in tables above, in order to study convergence theorems of nonlinear integrals in a unified way, it is better to separately consider the case where $\{f_n\}_{n\in\mathbb{N}}$ converges to f pointwise and the case where $\{f_n\}_{n\in\mathbb{N}}$ converges to f in measure. In addition, unlike the case of the Lebesgue integral, the monotone decreasing convergence theorem cannot directly be deduced from the monotone increasing convergence theorem because of the nonlinearity of integrals.

In the case of pointwise convergence the following functional forms of the monotone convergence theorem are fundamental and of wide application.

Theorem 5.1 (The monotone increasing convergence theorem) *Let* $I: \mathscr{M}(X) \times \mathscr{F}^+(X) \to [0, \infty]$ *be an integral functional and* $\mu \in \mathscr{M}(X)$. *Consider the following two conditions:*

(i) μ *is continuous from below.*
(ii) *The monotone increasing convergence theorem holds for* I_μ, *that is, for any* $\{f_n\}_{n\in\mathbb{N}} \subset \mathscr{F}^+(X)$ *and* $f \in \mathscr{F}^+(X)$, *if* $f_n \uparrow f$, *then* $I_\mu(f_n) \to I_\mu(f)$.

(1) *If* I_μ *is upper marginal continuous, measure-truncated, elementary, and perturbative with generator of continuous type, then (i) implies (ii).*
(2) *If* I_μ *is generative with limit preserving generator, then (ii) implies (i).*

An alternative form of the monotone increasing convergence theorem can be obtained by using the inner regularity of integral functionals.

Theorem 5.2 (An alternative form) *Let* $I: \mathscr{M}(X) \times \mathscr{F}^+(X) \to [0, \infty]$ *be an integral functional and* $\mu \in \mathscr{M}(X)$. *Consider the following two conditions:*

(i) μ *is continuous from below and* I_μ *is inner regular.*
(ii) *The monotone increasing convergence theorem holds for* I_μ, *that is, for any* $\{f_n\}_{n\in\mathbb{N}} \subset \mathscr{F}^+(X)$ *and* $f \in \mathscr{F}^+(X)$, *if* $f_n \uparrow f$, *then* $I_\mu(f_n) \to I_\mu(f)$.

(1) *If* I_μ *is elementary with generator of continuous type and pseudo-difference* $\ominus$ *is continuous on the triangular domain* $T := \{(a, b) \in [0, \infty]^2 : a > b\}$, *then (i) implies (ii).*
(2) *If* I_μ *is generative with limit preserving generator, then (ii) implies (i).*

The concept of uniform truncation of a family of functions is necessary to consider the monotone decreasing convergence theorem regardless of the types of nonlinear integrals.

Definition 5.10 Let $I: \mathscr{M}(X) \times \mathscr{F}^+(X) \to [0, \infty]$ be an integral functional and $\mu \in \mathscr{M}(X)$. A nonempty family $\mathscr{F} \subset \mathscr{F}^+(X)$ is called *uniformly truncated* for I_μ if for every $\varepsilon > 0$, there is a constant $c > 0$ such that

$$I_\mu(f) \le I_\mu(f \wedge c) + \varepsilon$$

for every $f \in \mathscr{F}$. A function $f \in \mathscr{F}^+(X)$ is called *truncated* for I_μ if the family $\mathscr{F} = \{f\}$ containing only f is uniformly truncated for I_μ.

Theorem 5.3 (The monotone decreasing convergence theorem) *Let* $I: \mathscr{M}(X) \times \mathscr{F}^+(X) \to [0, \infty]$ *be an integral functional and* $\mu \in \mathscr{M}(X)$. *Consider the following two conditions:*

(i) μ *is conditionally continuous from above.*
(ii) *The monotone decreasing convergence theorem holds for* I_μ, *that is, for any* $\{f_n\}_{n\in\mathbb{N}} \subset \mathscr{F}^+(X)$ *and* $f \in \mathscr{F}^+(X)$, *if* $f_n \downarrow f$, $I_\mu(f_1) < \infty$, *and* $\{f_n\}_{n\in\mathbb{N}}$ *is uniformly truncated for* I_μ, *then* $I_\mu(f_n) \to I_\mu(f)$.

(1) *If μ is finite and I_μ is elementary and perturbative with generator of continuous type, then (i) implies (ii).*
(2) *If I_μ is upper marginal continuous and generative with limit preserving generator of finite type, then (ii) implies (i).*

Remark 5.5 The uniform truncation of $\{f_n\}_{n\in\mathbb{N}}$ for I_μ in (ii) of Theorem 5.3 cannot be removed for the validity of monotone decreasing convergence theorems. In fact, if I_μ is upper marginal continuous, then for any decreasing sequence $\{f_n\}_{n\in\mathbb{N}} \subset \mathscr{F}^+(X)$ and $f \in \mathscr{F}^+(X)$, if $I_\mu(f_1) < \infty$ and $I_\mu(f_n) \to I_\mu(f)$, then $\{f_n\}_{n\in\mathbb{N}}$ is always uniformly truncated for I_μ.

Remark 5.6 In general, if $\mu \in \mathscr{M}(X)$ is null-additive, that is, $\mu(A \cup B) = \mu(A)$ whenever $A, B \in \mathscr{A}$ and $\mu(B) = 0$ and I_μ is s-monotone, then almost everywhere consistency holds for I_μ, that is, $I_\mu(f) = I_\mu(g)$ whenever $f, g \in \mathscr{F}^+(X)$ and $f = g$ μ-a.e. Therefore, if the null-additivity of μ and the s-monotonicity of I_μ are additionally assumed in Theorems 5.1, 5.2, and 5.3, then pointwise convergence of f_n to f may be replaced with almost everywhere convergence.

By Propositions 4.1, 4.2, 4.3, 4.4, and 4.5, the integral functionals $I =$ Ch, Si, Su, Sh are upper marginal continuous, measure-truncated, generative, elementary, and perturbative. In addition, their generators $\theta(a, b) = a \cdot b, a \wedge b$ are limit preserving and of finite and continuous type. Therefore, the following corollaries follow from Theorems 5.1 and 5.3. In fact, assumptions on f_n and μ in Corollaries 5.1 and 5.2 below are used for assuring conditions (i) and (ii) in Theorems 5.1 and 5.3 and reducing proofs to the case where μ is finite.

Corollary 5.1 (The monotone increasing convergence theorem) *Let $I =$* Ch, Si, Su, Sh. *Let $\mu \in \mathscr{M}(X)$, $\{f_n\}_{n\in\mathbb{N}} \subset \mathscr{F}^+(X)$, and $f \in \mathscr{F}^+(X)$. Assume that μ is continuous from below and $f_n \uparrow f$. Then $I_\mu(f_n) \to I_\mu(f)$.*

Corollary 5.2 (The monotone decreasing convergence theorem) *Let $\mu \in \mathscr{M}(X)$, $\{f_n\}_{n\in\mathbb{N}} \subset \mathscr{F}^+(X)$, and $f \in \mathscr{F}^+(X)$. Assume that μ is conditionally continuous from above and $f_n \downarrow f$.*

(1) Let $I =$ Ch, Si. *If $I_\mu(f_1) < \infty$, then $I_\mu(f_n) \to I_\mu(f)$.*
(2) If $\mu(\{f_1 > \mathrm{Su}(\mu, f)\}) < \infty$ (in particular, μ is finite), then $\mathrm{Su}_\mu(f_n) \to \mathrm{Su}_\mu(f)$.
(3) If $\mu(\{f_1 > 0\}) < \infty$ (in particular, μ is finite) and f_1 is μ-essentially bounded, then $\mathrm{Sh}_\mu(f_n) \to \mathrm{Sh}_\mu(f)$.

As is the case of ordinary measure theory, the monotone increasing and decreasing convergence theorems imply the Fatou lemma and the dominated convergence theorem.

Corollary 5.3 (The Fatou lemma) *Let $I =$* Ch, Si, Su, Sh. *Let $\mu \in \mathscr{M}(X)$ and let $\{f_n\}_{n\in\mathbb{N}} \subset \mathscr{F}^+(X)$ and $f \in \mathscr{F}^+(X)$. Assume that μ is continuous from below and $f_n \to f$. Then $I_\mu(f) \le \liminf_{n\to\infty} I_\mu(f_n)$.*

Corollary 5.4 (The dominated convergence theorem) *Let $\mu \in \mathscr{M}(X)$ and let $\{f_n\}_{n\in\mathbb{N}} \subset \mathscr{F}^+(X)$ and $f \in \mathscr{F}^+(X)$. Assume that μ is conditionally continuous and $f_n \to f$.*

(1) Let $I = \mathrm{Ch}, \mathrm{Si}$. If there is $g \in \mathscr{F}^+(X)$ such that $I_\mu(g) < \infty$ and $f_n \leq g$ for all $n \in \mathbb{N}$, then $I_\mu(f_n) \to I_\mu(f)$.
(2) If $\mu(\{\sup_{n\in\mathbb{N}} f_n > \mathrm{Su}(\mu, f)\}) < \infty$ (in particular, μ is finite), then $\mathrm{Su}_\mu(f_n) \to \mathrm{Su}_\mu(f)$.
(3) If $\mu(\{\sup_{n\in\mathbb{N}} f_n > 0\}) < \infty$ and $\sup_{n\in\mathbb{N}} f_n$ is μ-essentially bounded (in particular, there is a μ-essentially bounded $g \in \mathscr{F}^+(X)$ such that $\mu(\{g > 0\}) < \infty$ and $f_n \leq g$ for all $n \in \mathbb{N}$), then $\mathrm{Sh}_\mu(f_n) \to \mathrm{Sh}_\mu(f)$.

Remark 5.7 Since $I = \mathrm{Ch}, \mathrm{Si}, \mathrm{Su}, \mathrm{Sh}$ are s-monotone, if the null-additivity of μ is additionally assumed, then pointwise convergence of f_n to f may be replaced with almost everywhere convergence and condition $f_n \leq g$ may be replaced with $f_n \leq g$ μ-a.e. in Corollaries 5.1, 5.2, 5.3, and 5.4; see Remark 5.6.

The autocontinuity of nonadditive measures is required to establish convergence theorems for a sequence of measurable functions converging in measure.

Definition 5.11 (Wang [39]) Let $\mu \in \mathscr{M}(X)$.

(1) μ is called *autocontinuous from above* if $\mu(A \cup B_n) \to \mu(A)$ whenever $A \in \mathscr{A}$, $\{B_n\}_{n\in\mathbb{N}} \subset \mathscr{A}$, and $\mu(B_n) \to 0$.
(2) μ is called *autocontinuous from below* if $\mu(A \setminus B_n) \to \mu(A)$ whenever $A \in \mathscr{A}$, $\{B_n\}_{n\in\mathbb{N}} \subset \mathscr{A}$, and $\mu(B_n) \to 0$.
(3) μ is called *autocontinuous* if it is autocontinuous from above and below.

Example 5.2 The following nonadditive measures are autocontinuous.

(1) Every subadditive nonadditive measure and every nonadditive measure satisfying $\inf\{\mu(A) \colon A \in \mathscr{A}, A \neq \emptyset\} > 0$ are autocontinuous, but are not conditionally continuous in general.
(2) Let $m \colon \mathscr{A} \to [0, \infty]$ be a finitely additive measure. Let $\theta \colon [0, m(X)] \to [0, \infty]$ be an increasing function with $\theta(0) = 0$. Then the distorted measure

$$\mu(A) := \theta(m(A)), \quad A \in \mathscr{A},$$

determined by m and θ is autocontinuous if θ is continuous on $[0, m(X)]$ and strictly increasing on a neighborhood of the origin. In addition, μ is conditionally continuous if m is σ-additive and $\theta(\infty) = \infty$. In particular, the distorted measure $\mu(A) := m(A)^2 + \sqrt{m(A)}$ is autocontinuous, but neither subadditive nor superadditive. It is conditionally continuous if m is σ-additive.

Theorem 5.4 (A prototype of the Vitali convergence theorem) *Let $I \colon \mathscr{M}(X) \times \mathscr{F}^+(X) \to [0, \infty]$ be an integral functional and $\mu \in \mathscr{M}(X)$. Consider the following two conditions:*

(i) *μ is autocontinuous.*

(ii) *For any* $\{f_n\}_{n\in\mathbb{N}} \subset \mathscr{F}_0^+(X)$ *and* $f \in \mathscr{F}_0^+(X)$, *if* $f_n \xrightarrow{\mu} f$ *and* $\{f_n, f\}_{n\in\mathbb{N}}$ *is uniformly truncated for* I_μ, *then* $I_\mu(f_n) \to I_\mu(f)$.

(1) *If* μ *is finite and* I_μ *is perturbative, then (i) implies (ii).*
(2) *If* I_μ *is generative with limit preserving generator, then (ii) implies (i).*

As a matter of fact, Theorem 5.4 follows from the Fatou lemma and the reverse Fatou lemma below.

Theorem 5.5 (The Fatou lemma) *Let* $I : \mathscr{M}(X) \times \mathscr{F}^+(X) \to [0, \infty]$ *be an integral functional and* $\mu \in \mathscr{M}(X)$. *Consider the following two conditions:*

(i) μ *is autocontinuous from below.*
(ii) *The Fatou lemma holds for* I_μ, *that is, for any* $\{f_n\}_{n\in\mathbb{N}} \subset \mathscr{F}_0^+(X)$ *and* $f \in \mathscr{F}_0^+(X)$, *if* $f_n \xrightarrow{\mu} f$ *and* f *is truncated for* I_μ, *then*

$$I_\mu(f) \leq \liminf_{n\to\infty} I_\mu(f_n).$$

(1) *If* μ *is finite and* I_μ *is perturbative, then (i) implies (ii).*
(2) *If* I_μ *is generative with limit preserving generator, then (ii) implies (i).*

Theorem 5.6 (The reverse Fatou lemma) *Let* $I : \mathscr{M}(X) \times \mathscr{F}^+(X) \to [0, \infty]$ *be an integral functional and* $\mu \in \mathscr{M}(X)$. *Consider the following two conditions:*

(i) μ *is autocontinuous from above.*
(ii) *The reverse Fatou lemma holds for* I_μ, *that is, for any* $\{f_n\}_{n\in\mathbb{N}} \subset \mathscr{F}_0^+(X)$ *and* $f \in \mathscr{F}_0^+(X)$, *if* $f_n \xrightarrow{\mu} f$ *and* $\{f_n\}_{n\in\mathbb{N}}$ *is uniformly truncated for* I_μ, *then*

$$\limsup_{n\to\infty} I_\mu(f_n) \leq I_\mu(f).$$

(1) *If* μ *is finite and* I_μ *is perturbative, then (i) implies (ii).*
(2) *If* I_μ *is generative with limit preserving generator, then (ii) implies (i).*

The assumption μ being finite and f being truncated in the Fatou lemma (Theorem 5.5) can be removed by imparting the perturbation of not I_μ but I and further adding the upper marginal continuity and the measure-truncation of I.

Corollary 5.5 (The Fatou lemma) *Let* $I : \mathscr{M}(X) \times \mathscr{F}^+(X) \to [0, \infty]$ *be an integral functional and* $\mu \in \mathscr{M}(X)$. *If* μ *is autocontinuous from below and* I *is upper marginal continuous, measure-truncated, and perturbative, then the Fatou lemma holds for* I_μ, *that is, for any* $\{f_n\}_{n\in\mathbb{N}} \subset \mathscr{F}_0^+(X)$ *and* $f \in \mathscr{F}_0^+(X)$, *if* $f_n \xrightarrow{\mu} f$, *then* $I_\mu(f) \leq \liminf_{n\to\infty} I_\mu(f_n)$.

In order to formulate the Vitali convergence theorem and the bounded convergence theorem, the notions of uniform integrability and uniformly essential boundedness of measurable functions are required.

Definition 5.12 Let $I : \mathscr{M}(X) \times \mathscr{F}^{+}(X) \to [0, \infty]$ be an integral functional and $\mu \in \mathscr{M}(X)$. Let $\mathscr{F} \subset \mathscr{F}(X)$ be nonempty.

(1) $\mathscr{F}$ is called *uniformly integrable* for I_μ if

$$\lim_{c \to \infty} \sup_{f \in \mathscr{F}} I_\mu\big(\chi_{\{|f|>c\}}|f|\big) = 0.$$

(2) $\mathscr{F}$ is called *uniformly μ-essentially bounded* if there is a constant $c > 0$ such that

$$\mu(\{f \geq c\}) = 0 \text{ and } \mu(\{f \geq -c\}) = \mu(X)$$

for every $f \in \mathscr{F}$.

The Vitali convergence theorem and the bounded convergence theorem can be obtained as corollaries to the prototype of the Vitali convergence theorem (Theorem 5.4).

Corollary 5.6 (The Vitali convergence theorem) *Let $I : \mathscr{M}(X) \times \mathscr{F}^{+}(X) \to [0, \infty]$ be an integral functional and $\mu \in \mathscr{M}(X)$. Consider the following two conditions:*

(i) *μ is autocontinuous.*
(ii) *The Vitali convergence theorem holds for I_μ, that is, for any $\{f_n\}_{n\in\mathbb{N}} \subset \mathscr{F}_0^{+}(X)$ and $f \in \mathscr{F}_0^{+}(X)$, if $\{f_n\}_{n\in\mathbb{N}}$ is uniformly integrable for I_μ and $f_n \xrightarrow{\mu} f$, then $I_\mu(f) < \infty$ and $I_\mu(f_n) \to I_\mu(f)$.*

(1) *If μ is finite and I_μ is upper marginal continuous, horizontally subadditive, perturbative, and $I_\mu(r) < \infty$ for every $r > 0$, then (i) implies (ii).*
(2) *If I_μ is generative with limit preserving generator, then (ii) implies (i).*

Corollary 5.7 (The bounded convergence theorem) *Let $I : \mathscr{M}(X) \times \mathscr{F}^{+}(X) \to [0, \infty]$ be an integral functional and $\mu \in \mathscr{M}(X)$. Consider the following two conditions:*

(i) *μ is autocontinuous.*
(ii) *The bounded convergence theorem holds for I_μ, that is, for any $\{f_n\}_{n\in\mathbb{N}} \subset \mathscr{F}_0^{+}(X)$ and $f \in \mathscr{F}_0^{+}(X)$, if $\{f_n\}_{n\in\mathbb{N}}$ is uniformly μ-essentially bounded and $f_n \xrightarrow{\mu} f$, then f is μ-essentially bounded and $I_\mu(f_n) \to I_\mu(f)$.*

(1) *If μ is finite and I_μ is perturbative, then (i) implies (ii).*
(2) *If I_μ is generative with limit preserving generator, then (ii) implies (i).*

By Propositions 4.1, 4.2, 4.3, 4.4, and 4.5, the integral functionals $I =$ Ch, Si, Su, Sh are upper marginal continuous, horizontally subadditive, generative, and perturbative with limit preserving generator. In addition, $\mathrm{Su}_\mu(r) < \infty$ for every $r > 0$, and if μ is finite, then $I_\mu(r) < \infty$ for every $r > 0$ even in the case of $I =$ Ch, Si, Sh. Therefore, the following convergence theorems can be obtained.

Once again, assumptions on f_n, f and μ in Corollaries 5.8, 5.9, and 5.10 below are used for assuring conditions (i) and (ii) in the prototype of the Vitali convergence theorem (Theorem 5.4) and Corollaries 5.6 and 5.7 and reducing proofs to the case where μ is finite.

Corollary 5.8 (The Vitali convergence theorem for I = Ch, Si, Sh) *Let* I = Ch, Si, Sh. *Let* $\mu \in \mathcal{M}(X)$, $\{f_n\}_{n\in\mathbb{N}} \subset \mathcal{F}_0^+(X)$, *and* $f \in \mathcal{F}_0^+(X)$. *Assume that* μ *is finite and autocontinuous. If* $\{f_n\}_{n\in\mathbb{N}}$ *is uniformly integrable for* I_μ *and* $f_n \xrightarrow{\mu} f$, *then* $I_\mu(f) < \infty$ *and* $I_\mu(f_n) \to I_\mu(f)$.

Corollary 5.9 (The Vitali convergence theorem for I = Su) *Let* $\mu \in \mathcal{M}(X)$, $\{f_n\}_{n\in\mathbb{N}} \subset \mathcal{F}_0^+(X)$, *and* $f \in \mathcal{F}_0^+(X)$. *Assume that* μ *is autocontinuous. If* $f_n \xrightarrow{\mu} f$, *then* $\mathrm{Su}_\mu(f_n) \to \mathrm{Su}_\mu(f)$.

Remark 5.8 Neither the finiteness of μ nor the uniform integrability of $\{f_n\}_{n\in\mathbb{N}}$ is necessary in Corollary 5.9. This fact is due to a close relationship between the Sugeno integral and the Ky Fan metric [8] defined by

$$K(f, g) := \inf\{\varepsilon \in [0, \infty] \colon \mu(\{|f - g| > \varepsilon\}) \le \varepsilon\}, \quad f, g \in \mathcal{F}_0(X).$$

Since the Ky Fan metric characterizes convergence in μ-measure of measurable functions when μ is a σ-additive finite measure, the convergence theorems of the Sugeno integral have good compatibility with convergence in measure. In fact, for any $\mu \in \mathcal{M}(X)$, $\{f_n\}_{n\in\mathbb{N}} \subset \mathcal{F}_0(X)$, and $f \in \mathcal{F}_0(X)$, we have

$$K(f_n, f) \to 0 \Leftrightarrow f_n \xrightarrow{\mu} f \Leftrightarrow \mathrm{Su}_\mu(|f_n - f|) \to 0.$$

However, $\mathrm{Su}_\mu(f_n) \to \mathrm{Su}_\mu(f)$ cannot be derived from $\mathrm{Su}_\mu(|f_n - f|) \to 0$ unless μ is autocontinuous since the inequality

$$|\mathrm{Su}_\mu(f_n) - \mathrm{Su}_\mu(f)| \le \mathrm{Su}_\mu(|f_n - f|)$$

does not always hold because of the nonlinearity of the Sugeno integral.

The following bounded convergence theorem has already been discussed in [26] for the Choquet integral.

Corollary 5.10 (The bounded convergence theorem) *Let* I = Ch, Si, Sh. *Let* $\mu \in \mathcal{M}(X)$, $\{f_n\}_{n\in\mathbb{N}} \subset \mathcal{F}_0^+(X)$, *and* $f \in \mathcal{F}_0^+(X)$. *Assume that* μ *is finite and autocontinuous. If* $\{f_n\}_{n\in\mathbb{N}}$ *is uniformly* μ*-essentially bounded and* $f_n \xrightarrow{\mu} f$, *then* f *is* μ*-essentially bounded and* $I_\mu(f_n) \to I_\mu(f)$.

Remark 5.9 Recall that $\mu \in \mathcal{M}(X)$ is *strongly order continuous* if $\mu(A_n) \to 0$ whenever $\{A_n\}_{n\in\mathbb{N}} \subset \mathcal{A}$, $A \in \mathcal{A}$, $A_n \downarrow A$, and $\mu(A) = 0$. This condition was discovered in [25] as a necessary and sufficient condition for the validity of the Lebesgue theorem which states that every μ-almost everywhere convergent sequence of measurable functions converges in μ-measure. Since

$I = \mathrm{Ch}, \mathrm{Si}, \mathrm{Su}, \mathrm{Sh}$ are s-monotone and every autocontinuous nonadditive measure is null-additive, if the strong order continuity of μ is additionally assumed in Corollaries 5.8, 5.9, and 5.10, then by the Lebesgue theorem above and Remark 5.6, convergence of f_n to f in μ-measure may be replaced with μ-almost everywhere convergence.

Every integral functional $I\colon \mathscr{M}(X) \times \mathscr{F}^+(X) \to [0, \infty]$ can be extended in the following two ways

$$I^s(\mu, f) := I(\mu, f^+) - I(\mu, f^-), \quad (\mu, f) \in \mathscr{M}(X) \times \mathscr{F}(X),$$

$$I^a(\mu, f) := I(\mu, f^+) - I(\bar{\mu}, f^-), \quad (\mu, f) \in \mathscr{M}_b(X) \times \mathscr{F}(X)$$

in order to consider the integrals of not necessarily nonnegative functions. The functional I^s is called the *symmetric extension* of I or the *symmetric integral* determined by I, while I^a is called the *asymmetric extension* of I or the *asymmetric integral* determined by I. They are not defined if the right-hand side is of the form $\infty - \infty$. The symmetric integral I^s is symmetric in the sense that

$$I^s(\mu, -f) = -I^s(\mu, f)$$

and the asymmetric integral I^a is asymmetric in the sense that

$$I^a(\mu, -f) = -I^a(\bar{\mu}, f).$$

Although details are omitted, all results in this section hold for the symmetric integral I^s and the asymmetric integral I^a with appropriate modifications.

6 Concluding Remarks

In this article, some of the distribution-based integrals, such as the Choquet, Šipoš, Sugeno, and Shilkret integrals, are introduced. Then, by considering a nonlinear integral as a nonlinear functional defined on an appropriate domain, the properties common to those integrals are summarized from a functional analytic viewpoint (Sects. 3 and 4). Among those properties, the perturbation of nonlinear functionals plays an essential role in our unified approach to convergence theorems of nonlinear integrals. In fact, it allows us to discuss in a unified way the previous and best results known as the convergence theorems of nonlinear integrals (Sect. 5). This approach also has sufficient applicability to the studies of monotone convergence theorems for a net of semicontinuous functions [18] and weak convergence of nonadditive measures (for instance, nonadditive portmanteau theorem and the uniformity of weak convergence) [15].

Unlike the Choquet integral, the Šipoš integral can be defined independently of the Lebesgue integral. Moreover, it coincides with the abstract Lebesgue integral for any σ-additive measure, so that the convergence theorems introduced in Sect. 5 also hold for the abstract Lebesgue integral. These advantages suggest that the theory of the present Lebesgue integral, at least the theory leading to convergence theorems, can simultaneously be developed for both the abstract Lebesgue and Choquet integrals if it will be reconstructed by using the Šipoš integral.

In addition to distribution-based integrals, there are also decomposition-based integrals such as the pan integral and the concave integral. The investigation of the convergence of those nonlinear integrals is a future task.

If you are interested in the theory of nonadditive measures and nonlinear integrals, the author would appreciate it if you could see another expository article [16].

Acknowledgements This work was supported by JSPS KAKENHI Grant Number 17K05293.

References

1. Benvenuti P, Mesiar R, Vivona D (2002) Monotone set functions-based integrals. In: Pap E (ed) Handbook of measure theory. Elsevier, Amsterdam, pp 1329–1379
2. Berberian SK (1962) Measure and integration. Chelsea Publishing Company, New York
3. Choquet G (1953–1954) Theory of capacities. Ann Inst Fourier (Grenoble) 5:131–295
4. Denneberg D (1997) Non-additive measure and integral, 2nd edn. Kluwer Academic Publishers, Dordrecht
5. Dobrakov I (1974) On submeasures I. Dissertationes Math (Rozprawy Mat) 112:1–35
6. Ellsberg D (1961) Risk, ambiguity, and the savage axioms. Quart J Econ 75:643–669
7. Even Y, Lehrer E (2014) Decomposition-integral: unifying Choquet and the concave integrals. Econ Theory 56:33–58
8. Fan Ky (1944) Entfernung zweier zufälligen Grössen und die Konvergenz nach Wahrscheinlichkeit. Math Zeitschr 49:681–683
9. Grabisch M (2016) Set functions, games and capacities in decision making. Springer, Cham
10. Grabisch M, Murofushi T, Sugeno M (eds) (2000) Fuzzy measures and integrals, theory and applications. Physica-Verlag, Heidelberg
11. Halmos PR (1974) Measure theory. Springer, New York
12. Kawabe J (2012) Metrizability of the Lévy topology on the space of nonadditive measures on metric spaces. Fuzzy Sets Syst 204:93–105
13. Kawabe J (2014) Weak convergence of nonadditive measures defined by Choquet and Sugeno integrals. In: Kato M, Maligranda L (eds) Banach and function spaces IV, Yokohama Publishers, Yokohama, pp 63–79
14. Kawabe J (2015) The bounded convergence in measure theorem for nonlinear integral functionals. Fuzzy Sets Syst 271:31–42
15. Kawabe J (2016) Weak convergence of nonadditive measures based on nonlinear integral functionals. Fuzzy Sets Syst 289:1–15
16. Kawabe J (2016) Nonadditive measures and nonlinear integrals. Sugaku 68:266–292 (in Japanese) English translation will be appeared in Sugaku Expositions published by American Mathematical Society
17. Kawabe J (2016) A unified approach to the monotone convergence theorem for nonlinear integrals. Fuzzy Sets Syst 304:1–19

18. Kawabe J (2016) The monotone convergence theorems for nonlinear integrals on a topological space. Linear Nonlinear Anal 2:281–300
19. Kawabe J (2017) The Vitali type theorem for the Choquet integral. Linear Nonlinear Anal 3:349–365
20. Kawabe J (Submitted) The Vitali convergence theorem for nonlinear integral functionals
21. Lehrer E (2009) A new integral for capacities. Econ Theory 39:157–176
22. Lehrer E, Teper R (2008) The concave integral over large spaces. Fuzzy Sets Syst 159:2130–2144
23. Levi B (1906) Sopra l'integrazione delle serie. Rend Istituto Lombardo di Sci e Lett (Ser. 2) 39:775–780
24. Levy H (1992) Stochastic dominance and expected utility: survey and analysis. Manag Sci 38:555–593
25. Li J (2003) Order continuous of monotone set function and convergence of measurable functions sequence. Appl Math Comput 135:211–218
26. Murofushi T, Sugeno M, Suzaki M (1997) Autocontinuity, convergence in measure, and convergence in distribution. Fuzzy Sets Syst 92:197–203
27. Nishimura KG, Ozaki H (2017) Economics of pessimism and optimism. Springer, Japan
28. Pap E (1995) Null-additive set functions. Kluwer Academic Publishers, Bratislava
29. Rao MM (2004) Measure theory and integration, 2nd edn, revised and expanded. Marcel Dekker, New York
30. Ralescu D, Adams G (1980) The fuzzy integral. J Math Anal Appl 75:562–570
31. Schmeidler D (1986) Integral representation without additivity. Proc Am Math Soc 97:255–261
32. Shilkret N (1971) Maxitive measure and integration. Indag Math 33:109–116
33. Šipoš J (1979) Integral with respect to a pre-measure. Math Slovaca 29:141–155
34. Šipoš J (1979) Non linear integrals. Math Slovaca 29:257–270
35. Song J, Li J (2005) Lebesgue theorems in non-additive measure theory. Fuzzy Sets Syst 149:543–548
36. Sugeno M (1974) Theory of fuzzy integrals and its applications. Doctoral thesis, Tokyo Inst of Tech, Tokyo
37. Sugeno M, Murofushi T (1987) Pseudo-additive measures and integrals. J Math Anal Appl 122:197–222
38. Vitali G (1907) Sull'integrazione per serie. Rend del Circolo Mat di Palermo 23:137–155
39. Wang Z (1984) The autocontinuity of set function and the fuzzy integral. J Math Anal Appl 99:195–218
40. Wang Z (1997) Convergence theorems for sequences of Choquet integrals. Int J Gen Syst 26:133–143
41. Wang Z, Klir GJ (1992) Fuzzy measure theory. Plenum Press, New York
42. Wang Z, Klir GJ (2009) Generalized measure theory. Springer, New York
43. Yang Q (1985) The pan-integral on the fuzzy measure space. J Fuzzy Math 4:107–114 (in Chinese)
44. Zhao RH (1981) (N) fuzzy integral. J Math Res Exposition 1:55–72 (in Chinese)
45. Zhao RH (1985) Continuity and the Fubini theorem for (N) fuzzy integrals. J Xinjiang Univ Nat Sci 2:95–106 (in Chinese)

A Two-Sector Growth Model with Credit Market Imperfections and Production Externalities

Takuma Kunieda and Kazuo Nishimura

Abstract A two-sector dynamic general equilibrium model with financial constraints and production externalities is studied. Agents face idiosyncratic productivity shocks in each period. Agents who draw high productivity borrow resources in the financial market and become capital producers, whereas agents who draw low productivity become lenders. We analyze how the interaction between the extent of financial constraints and sector-specific production externalities affects the characterization of equilibria in a two-sector economy.

Keywords Two production sectors · Financial constraints · Production externalities · Indeterminacy

Article type: Research Article
Received: February 19, 2018
Revised: March 21, 2018

JEL Classification: E32, E44, O41
Mathematics Subject Classification (2010): 37N40, 39A30

T. Kunieda
School of Economics, Kwansei Gakuin University, Nishinomiya, Hyogo, Japan
e-mail: tkunieda@kwansei.ac.jp

K. Nishimura (✉)
Research Institute of Economy, Trade and Industry, Tokyo, Japan
Research Institute for Economics and Business Administration, Kobe University, Kobe, Japan
e-mail: nishimura@rieb.kobe-u.ac.jp

S. Kusuoka, T. Maruyama (eds.), *Advances in Mathematical Economics*, Advances in Mathematical Economics 22, https://doi.org/10.1007/978-981-13-0605-1_5

1 Introduction

We investigate the local dynamics in the neighborhood of a steady state in an economy where agents face financial constraints and goods are produced in two sectors with sector-specific production externalities.

There are earlier works that investigate the local dynamics in an economy with financial constraints. Among others, Kiyotaki and Moore [16] and Cordoba and Ripoll [13, 14] demonstrate that equilibrium is unique and dampening cycles can occur in a collateral-constrained economy. In the model by Cordoba and Ripoll [14], a unique steady state is a saddle point under plausible parameter conditions. Woodford [24] and Barinci and Chéron [5] focus on an economy where capitalists and workers coexist and show that indeterminacy can occur because workers face financial constraints.[1] Benhabib and Wang [9] and Liu and Wang [18] create a mechanism where financial constraints are the potential source of indeterminacy. In their models, the presence of a fixed cost directly or indirectly affecting financial constraints is a crucial factor for indeterminacy to occur. All the abovementioned studies assume that an economy is endowed with an aggregate production sector in which only one final good is produced. In the current study, we consider an economy with two production sectors. We analyze how the interaction between the extent of financial constraints and the sector-specific production externalities affects the characterization of equilibria in a two-sector economy.[2]

Over the past 20 years, many researchers have investigated multiple equilibria or indeterminacy of equilibria in dynamic general equilibrium models.[3] It is well known that indeterminacy causes self-fulfilling sunspot business fluctuations (Shell [22]; Azariadis [4]; Cass and Shell [12]; Woodford [23]). In this literature, production externalities have been an important feature of the model because they are a source of inefficiency that causes indeterminacy of equilibria. Among others, Benhabib and Nishimura [8] demonstrate that indeterminacy can easily occur in a model with two production sectors even though production externalities are small, provided that capital goods are labor intensive from the private perspective and capital intensive from the social perspective. In the current paper, a financial market is explicitly introduced in the two-production-sector model. In particular, agents face financial constraints and can borrow only up to a certain proportion of their own funds in our model. In each period, they receive an idiosyncratic productivity shock. Agents who draw higher productivity become borrowers (capital producers), and agents who draw lower productivity become lenders. In other words, borrowers and lenders endogenously appear in equilibrium.

---

[1]In contrast to Woodford [24], Barinci and Chéron [5] consider an economy with increasing return-to-scale production externalities.

[2]We do not assume any fixed cost that affects financial constraints.

[3]See, for instance, Benhabib and Farmer [6, 7], Benhabib and Nishimura [8], Borldrin and Rustichini [11], Benhabib et al. [10], Nishimura and Shimomura [19], Nishimura and Venditti [20, 21], and Dufourt et al. [15].

Our main finding is that under the moderate parameter conditions, if financial constraints are severe, equilibrium is uniquely determined, with the steady state being a saddle point, whereas if the financial constraints are relaxed, equilibrium is indeterminate, with the steady state being totally stable.

The remainder of this paper proceeds as follows. In the next section, the model is presented in which the two production sectors exhibit sector-specific production externalities and agents face financial constraints. In Sect. 3, we derive equilibrium and the dynamical system. In Sect. 4, we characterize the dynamic property around the steady state and obtain the condition for the extent of financial constraints to produce multiple equilibria. Section 5 concludes the current study.

2 Model

A closed economy continues from time 0 to $+\infty$ in discrete time and consists of an infinitely lived representative firm and infinitely lived agents, whose population is equal to 1. In each period, the representative firm produces both consumption and intermediate goods. The intermediate goods are numeraire throughout the current analysis. The infinitely lived agents have potential investment opportunities to produce capital from the intermediate goods but receive uninsured idiosyncratic productivity shocks in each period that affect the productivity in capital production.

2.1 Agents

2.1.1 Timing of Events

Figure 1 illustrates the timing of the events at time t. At the beginning of time t when the idiosyncratic productivity shock has not yet been realized, an agent earns the following incomes: a wage income, returns to her saving, and a lump-sum profit from the representative firm. The consumption good market at time t is opened at the beginning of the period and is closed before the idiosyncratic productivity shock is realized. Therefore, she must make a decision about consumption and saving at the beginning of time t without knowing her productivity in order to produce capital used at time $t+1$. At the end of time t, the idiosyncratic productivity shock is realized. There are two saving methods: one is lending her savings in the financial market and the other is initiating an investment project. She optimally chooses one of the saving methods with knowing her productivity. Lending one unit of savings in the financial market at time t yields a claim to r_{t+1} units of intermediate goods at time $t+1$, where r_{t+1} is the gross real interest rate. Purchasing one unit of intermediate goods at time t for capital production creates Φ_t units of capital used at time $t+1$, which are sold at price, q_{t+1}, to the production sectors at time $t+1$. Although she can borrow in the financial market when she starts to produce capital, she faces a financial constraint and can borrow up to a certain proportion of her own funds.

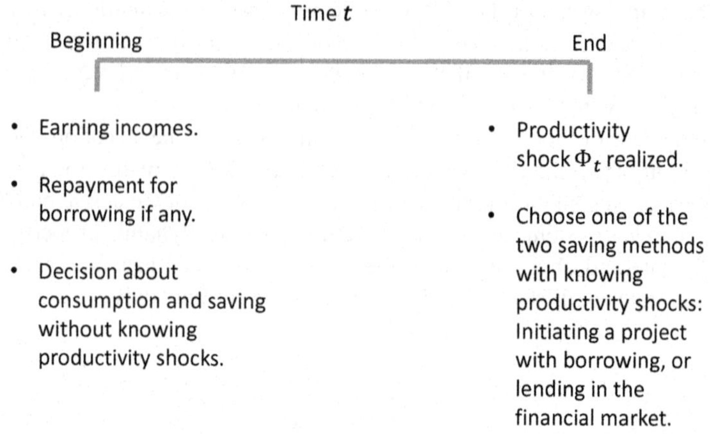
Time t
Beginning
End
Earning incomes.
Repayment for borrowing if any.
Decision about consumption and saving without knowing productivity shocks.
Productivity shock Φ_t realized.
Choose one of the two saving methods with knowing productivity shocks: Initiating a project with borrowing, or lending in the financial market.

Fig. 1 Timing of events and an agent's behavior

Productivity Φ_t is a random variable on a probability space $(\Omega, \mathscr{F}, P)$, where Ω is a sample space (for which one can assume $\Omega = [0, 1]$), $\mathscr{F}$ is a σ-algebra on Ω, and P is the probability measure. Φ_t is a function of $\omega_t \in \Omega$, and its support is $[0, \eta]$, where $\eta \in (0, \infty)$. The cumulative distribution function of Φ_t is denoted by $G(\Phi) := P(\{\omega_t \in \Omega \mid \Phi_t(\omega_t) \leq \Phi\})$, which is time-invariant and continuously differentiable on the support, where $\{\omega_t \in \Omega \mid \Phi_t(\omega_t) \leq \Phi\} \in \mathscr{F}$. $\Phi_0, \Phi_1, \ldots,$ are independent and identically distributed across both agents and time (the i.i.d. assumption). There is no insurance market for the productivity shocks, and thus, no one can insure against low productivity. Denote the history of ω_t by $\omega^{t-1} = \{\omega_0, \omega_1, \ldots, \omega_{t-1}\}$. Then, we can define a probability space $(\Omega^t, \mathscr{F}^t, P^t)$ that is a Cartesian product of t copies of $(\Omega, \mathscr{F}, P)$, where ω^{t-1} is an element of Ω^t. Because the measure of the agent population is equal to one and because of the i.i.d. assumption, ω^{t-1} can denote an individual who experiences the history, $\omega^{t-1} = \{\omega_0, \omega_1, \ldots, \omega_{t-1}\}$.

2.1.2 Maximization Problem

An agent solves a maximization problem for her lifetime utility given in the following:

$$\max E\left[\sum_{\tau=t}^{\infty} \beta^{\tau-t} c_\tau(\omega^{\tau-1}) \middle| \omega^{t-1}\right]$$

subject to

$$p_\tau c_\tau(\omega^{\tau-1}) + s_\tau(\omega^{\tau-1}) =$$

$$q_\tau \Phi_{\tau-1}(\omega_{\tau-1}) x_{\tau-1}(\omega^{\tau-1}) + r_\tau b_{\tau-1}(\omega^{\tau-1}) + w_\tau + \pi_\tau \tag{1}$$

$$b_\tau(\omega^\tau) \geq -\lambda s_\tau(\omega^{\tau-1}) \tag{2}$$

$$x_\tau(\omega^\tau) \geq 0, \tag{3}$$

for $\tau \geq t$, where $\beta \in (0, 1)$ is the subjective discount factor, $c_\tau(\omega^{\tau-1})$ is consumption, and $E[.|\omega^{t-1}]$ is an expectation operator given the history, ω^{t-1}. In what follows, by using (2) and (3), we transform Eq. (1) into one budget constraint given by Eq. (6) below.

In (1), w_τ and π_τ are a wage income and a profit that is given from the representative firm in a lump-sum manner, respectively, and p_τ is the price of consumption goods. $s_\tau(\omega^{\tau-1}) := x_\tau(\omega^\tau) + b_\tau(\omega^\tau)$ is the agent's saving at time τ, where $x_\tau(\omega^\tau)$ is intermediate goods used for capital production and $b_\tau(\omega^\tau)$ is lending if $b_\tau(\omega^\tau) > 0$ and borrowing if $b_\tau(\omega^\tau) < 0$. A linear technology with respect to intermediate goods is assumed for capital production such as $\Phi_{\tau-1}(\omega_{\tau-1})x_{\tau-1}(\omega^{\tau-1})$, which is capital produced at time τ. Equation (1) implies that when the agent makes a decision at time t about consumption, $c_t(\omega^{t-1})$, and saving, $s_t(\omega^{t-1})$, she does not know her productivity, $\Phi_t(\omega_t)$, as previously discussed. However, as noted from the expression $s_t(\omega^{t-1}) = x_t(\omega^t) + b_t(\omega^t)$, she knows $\Phi_t(\omega_t)$ when she makes a portfolio decision about investing in a capital production project, lending, and/or borrowing at time t. Equation (1) is the flow budget constraint effective for $\tau \geq 1$. At time 0, the flow budget constraint is assumed to be given by $p_0 c_0 + s_0 = q_0 K_0 + w_0 + \pi_0$, where K_0 is the initial capital endowment that is common across agents.

Inequality (2) is the financial constraint the agent faces at time τ.[4] Even though an agent is willing to borrow in the financial market, she can do so only up to a partial proportion of her savings, which is her own fund. In (2), $\lambda \in (0, \infty)$ is the extent of financial constraints: the smaller λ is, the more severe the financial constraint is. Inequality (2) can be rewritten as $b_\tau(\omega^\tau) \geq -\mu x_\tau(\omega^\tau)$, where $\mu = \lambda/(1 + \lambda) \in (0, 1)$. Because this constraint is more convenient than inequality (2), we use it henceforth. As μ goes to 1, the financial market approaches perfection, and as μ goes to zero, agents are unable to borrow in the financial market. The purchase of intermediate goods should be nonnegative, and thus, inequality (3) is imposed.

2.1.3 Optimal Portfolio Decision Within a Period

We define $\phi_t := r_{t+1}/q_{t+1}$. With knowing the productivity in capital production, agents who draw $\Phi_t > \phi_t$ optimally borrow up to the limit of the financial constraint and purchase intermediate goods for capital production, whereas agents who draw $\Phi_t \leq \phi_t$ lend all their savings in the financial market to acquire the gross interest,

[4] This type of financial constraints is employed by many researchers such as Aghion et al. [2], Aghion and Banerjee [1], and Aghion et al. [3].

r_{t+1}.[5] Hence, ϕ_t is the cutoff for the productivity shocks that divide agents into lenders and borrowers (capital producers) at time t, and an agent's portfolio program is given by

$$x_t(\omega^t) = \begin{cases} 0 & \text{if } \Phi_t(\omega_t) \le \phi_t \\ \frac{s_t(\omega^{t-1})}{1-\mu} & \text{if } \Phi_t(\omega_t) > \phi_t, \end{cases} \tag{4}$$

and

$$b_t(\omega^t) = \begin{cases} s_t(\omega^{t-1}) & \text{if } \Phi_t(\omega_t) \le \phi_t \\ -\frac{\mu}{1-\mu} s_t(\omega^{t-1}) & \text{if } \Phi_t(\omega_t) > \phi_t. \end{cases} \tag{5}$$

2.1.4 Euler Equation

The portfolio program given by (4) and (5) rewrites the flow budget constraint (1) as

$$s_\tau(\omega^{\tau-1}) + p_\tau c_\tau(\omega^{\tau-1}) = R_\tau(\omega_{\tau-1}) s_{\tau-1}(\omega^{\tau-2}) + w_\tau + \pi_\tau, \tag{6}$$

where $R_\tau(\omega_{\tau-1}) := \max\{r_\tau, (q_\tau \Phi_{\tau-1}(\omega_{\tau-1}) - r_\tau \mu)/(1-\mu)\}$. The maximization of the agent's lifetime utility subject to (6) yields the Euler equation as follows:

$$p_{t+1} = \beta E\left[R_{t+1}(\omega_t)|\omega^{t-1}\right] p_t. \tag{7}$$

The necessary and sufficient optimality conditions for the lifetime utility maximization problem consist of the Euler equation (7) as well as the transversality condition $\lim_{\tau\to\infty} \beta^\tau E[s_{t+\tau}(\omega^{t+\tau-1})/p_{t+\tau}|\omega^{t-1}] = 0$.

2.2 *Production*

The representative firm produces both intermediate and consumption goods, being endowed with Cobb-Douglas technologies:

$$\bar{F}^1(l_t^1, k_t^1, \bar{l}_t^1, \bar{k}_t^1) = A(l_t^1)^{\alpha_L^1}(k_t^1)^{\alpha_K^1}(\bar{l}_t^1)^{a_L^1}(\bar{k}_t^1)^{a_K^1}$$

for intermediate goods, and

$$\bar{F}^2(l_t^2, k_t^2, \bar{l}_t^2, \bar{k}_t^2) = B(l_t^2)^{\alpha_L^2}(k_t^2)^{\alpha_K^2}(\bar{l}_t^2)^{a_L^2}(\bar{k}_t^2)^{a_K^2}$$

[5] The derivation of an optimal portfolio allocation of savings follows Kunieda and Shibata [17]. Although agents who draw $\Phi_t = \phi_t$ are indifferent between initiating a capital production project and lending in the financial market, it is assumed that they lend their savings in the financial market.

for consumption goods, where $\alpha_L^i, \alpha_K^i \in (0,1)$, $\alpha_L^i + \alpha_K^i + a_L^i + a_K^i = 1$ for $i = 1, 2$, and $\Delta := \alpha_L^1\alpha_K^2 - \alpha_L^2\alpha_K^1 \neq 0$. A and B are the productivity parameters. In the production functions, l_t^i and k_t^i are labor and capital used for the production of each good, respectively, and $\bar{l}_t^i$ and $\bar{k}_t^i$ are the components of production externalities with respect to labor and capital, respectively. In equilibrium, it holds that $l_t^i = \bar{l}_t^i$ and $k_t^i = \bar{k}_t^i$, although $\bar{l}_t^i$ and $\bar{k}_t^i$ are exogenous when the firm solves the profit maximization problem. The firm solves the following maximization problem:

$$\max_{l_t^1, l_t^2, k_t^1, k_t^2} \Pi_t := \bar{F}^1(l_t^1, k_t^1, \bar{l}_t^1, \bar{k}_t^1) + p_t\bar{F}^2(l_t^2, k_t^2, \bar{l}_t^2, \bar{k}_t^2) + (1-\delta)k_t - q_tk_t - w_tl_t, \tag{8}$$

where $\delta \in (0,1)$ is the capital depreciation rate and $k_t = k_t^1 + k_t^2$ is the total capital in the economy. It is assumed that the remaining capital, $(1-\delta)k_t$, after production at time t can be used as intermediate goods, and thus, the total intermediate goods, $\bar{F}^1(l_t^1, k_t^1, \bar{l}_t^1, \bar{k}_t^1) + (1-\delta)k_t$, are sold to capital producers. The total labor supply is given by $l_t^1 + l_t^2 = l_t$, which is equal to the population of agents, i.e., $l_t = 1$. The first-order conditions for the profit maximization problem are given by

$$A\alpha_L^1\left(\frac{k_t^1}{l_t^1}\right)^{1-\theta_1} = p_tB\alpha_L^2\left(\frac{k_t^2}{l_t^2}\right)^{1-\theta_2} = w_t. \tag{9}$$

and

$$A\alpha_K^1\left(\frac{k_t^1}{l_t^1}\right)^{-\theta_1} = p_tB\alpha_K^2\left(\frac{k_t^2}{l_t^2}\right)^{-\theta_2} = q_t + \delta - 1, \tag{10}$$

where $\theta_i := \alpha_L^i + a_L^i$ for $i = 1, 2$. Note that we have used equilibrium conditions, $l_t^i = \bar{l}_t^i$ and $k_t^i = \bar{k}_t^i$, to obtain (9) and (10). Assumption 2.1 below is imposed so that the law of demand for each input is satisfied.

Assumption 2.1 $\theta_i \in (0,1)$ *for* $i = 1, 2$.

Equations (9) and (10) yield

$$k_t^1 = \frac{\alpha_K^1 w_t}{\alpha_L^1(q_t+\delta-1)}l_t^1 \text{ and } k_t^2 = \frac{\alpha_K^2 w_t}{\alpha_L^2(q_t+\delta-1)}l_t^2 \tag{11}$$

Equations (9) and (10) also yield

$$w_t = \Psi p_t^{\frac{1-\theta_1}{\theta_2-\theta_1}} =: w(p_t) \text{ and } q_t + \delta - 1 = \Lambda p_t^{\frac{-\theta_1}{\theta_2-\theta_1}} =: v(p_t), \tag{12}$$

where

$$\Psi := [(A(\alpha_L^1)^{\theta_1}(\alpha_K^1)^{1-\theta_1})^{\theta_2-1}(B(\alpha_L^2)^{\theta_2}(\alpha_K^2)^{1-\theta_2})^{1-\theta_1}]^{1/(\theta_2-\theta_1)}$$

and

$$\Lambda := [(A(\alpha_L^1)^{\theta_1}(\alpha_K^1)^{1-\theta_1})^{\theta_2}(B(\alpha_L^2)^{\theta_2}(\alpha_K^2)^{1-\theta_2})^{-\theta_1}]^{1/(\theta_2-\theta_1)}.$$

As in Benhabib and Nishimura [8], it is said that if $\Delta = \alpha_L^1\alpha_K^2 - \alpha_L^2\alpha_K^1 > (<)0$, the intermediate goods are labor (capital) intensive from the private perspective and if $\theta_1 > (<)\theta_2$, the intermediate goods are labor (capital) intensive from the social perspective.

3 Equilibrium

A competitive equilibrium is expressed by sequences of prices $\{w_t, q_t, p_t, r_{t+1}\}$ for all $t \geq 0$ and allocation $\{k_t, k_t^1, k_t^2, l_t, l_t^1, l_t^2\}$ for all $t \geq 0$ and $\{c_t(\omega^{t-1}), s_t(\omega^{t-1}), x_t(\omega^t), b_t(\omega^t)\}$ for all $t \geq 0$, ω^t, and ω^{t-1}, so that (i) for each ω^t and ω^{t-1}, the consumer maximizes her lifetime utility from time t onward; (ii) the representative firm maximizes its profits in each period; and (iii) consumption and intermediate goods markets, a financial market, a capital market, and a labor market are clear.[6]

3.1 Market-Clearing Conditions

Because in each time, the total consumption is equal to the production of consumption goods, the consumption goods market-clearing condition is given by

$$C_t := \int_{\Omega^t} c_t(\omega^{t-1})dP^t(\omega^{t-1}) = F^2(k_t^2, l_t^2), \tag{13}$$

where $F^2(k_t^2, l_t^2) := \bar{F}^2(l_t^2, k_t^2, l_t^2, k_t^2)$. As seen in (4), the intermediate goods are purchased by agents who draw higher productivity, such that $\Phi_t(\omega_t) > \phi_t$. Therefore, the intermediate goods market-clearing condition is given by

$$\int_{\Omega^t\times(\Omega\backslash\Xi_t)} x_t(\omega^t)dP^{t+1}(\omega^t) = F^1(k_t^1, l_t^1) + (1-\delta)k_t, \tag{14}$$

where $\Xi_t = \{\omega_t \in \Omega | \Phi_t(\omega_t) \leq \phi_t\}$ and $F^1(k_t^1, l_t^1) := \bar{F}^1(l_t^1, k_t^1, l_t^1, k_t^1)$. Because, in the financial market, all lending and borrowing are canceled out, it follows that

$$\int_{\Omega^{t+1}} b_t(\omega^t)dP^{t+1}(\omega^t) = 0, \tag{15}$$

which is the financial market-clearing condition. From (15), we obtain the following lemma.

---

[6]To be accurate, c_0 is not subject to any history of the stochastic events and ω^{-1} is empty.

Lemma 3.1 *The financial market-clearing condition (15) is satisfied if and only if*

$$\int_{\Omega^t \times \Xi_t} s_t(\omega^{t-1}) dP^{t+1}(\omega^t) = \frac{\mu}{1-\mu} \int_{\Omega^t \times (\Omega \setminus \Xi_t)} s_t(\omega^{t-1}) dP^{t+1}(\omega^t).$$

Proof Inserting (5) into the financial market-clearing condition (15) yields the desired equation. □

Capital is supplied by agents who draw higher productivity such that $\Phi_t(\omega_t) > \phi_t$ and is demanded by the representative firm. Hence, the capital market-clearing condition is given by

$$k_{t+1}^1 + k_{t+1}^2 = \int_{\Omega^t \times (\Omega \setminus \Xi_t)} \Phi_t(\omega_t) x_t(\omega^t) dP^{t+1}(\omega^t). \tag{16}$$

Finally, the labor market-clearing condition is given by

$$l_t^1 + l_t^2 = l_t = 1. \tag{17}$$

3.2 *Production in Equilibrium*

From (11), $k_t^1 + k_t^2 = k_t$, and $l_t^1 + l_t^2 = 1$, the production functions become as follows:

$$F^1(l_t^1, k_t^1) = -\frac{\alpha_L^2 v(p_t) k_t - \alpha_K^2 w(p_t)}{\Delta} \tag{18}$$

and

$$p_t F^2(l_t^2, k_t^2) = \frac{\alpha_L^1 v(p_t) k_t - \alpha_K^1 w(p_t)}{\Delta}. \tag{19}$$

From (18) and (19), the gross product, $Y_t = F^1(l_t^1, k_t^1) + p_t F^2(l_t^2, k_t^2)$, is obtained as follows:

$$Y_t = \frac{(\alpha_L^1 - \alpha_L^2) v(p_t) k_t + (\alpha_K^2 - \alpha_K^1) w(p_t)}{\Delta}. \tag{20}$$

3.3 *Cutoff*

The financial market-clearing condition (15) yields a time-invariant cutoff, $\phi_t = \phi^*$, in equilibrium that divides agents into lenders and borrowers.

Proposition 3.1 *The cutoff, ϕ^*, is given by*

$$G(\phi^*) = \mu. \tag{21}$$

Proof See the Appendix.

Because the cumulative distribution function is strictly increasing over the support, $\phi^* = G^{-1}(\mu)$ is uniquely determined. As μ increases, ϕ^* increases. This means that as the financial constraints are relaxed, the number of lenders increases and the number of capital producers decreases. Although the number of capital producers decreases, allocative inefficiency with respect to the use of intermediate goods is corrected. This is because the intermediate goods are intensively used by more highly productive agents.

3.4 Dynamical System

To aggregate the flow budget constraint (6) across all agents, we obtain Lemma 3.2 below, which follows from the financial market-clearing condition (15).

Lemma 3.2

$$\int_{\Omega^t} R_t(\omega_{t-1}) s_{t-1}(\omega^{t-2}) dP^t(\omega^{t-1}) = q_t k_t \tag{22}$$

Proof See the Appendix.

Lemma 3.2 implies that capital producers sell capital to the representative firm at price q_t. From the microeconomic perspective, the savings of agents who draw lower productivity are lent out to agents who draw higher productivity, and the lenders are repaid with interest. Therefore, although the returns to individual savings vary depending upon the individual productivity, the total income from all agents' savings is eventually equal to the value of total capital in the economy.

The use of Lemma 3.2 aggregates the flow budget constraint (6) across all agents and obtains the relationship between the total demand for and the total supply of intermediate goods. The total funds available for capital production consist of capital producers' savings and their borrowing from lenders through the financial market, which is equal to the aggregate saving across all agents. The total funds are used to purchase the intermediate goods. Lemma 3.3 below describes this situation.

Lemma 3.3

$$\int_{\Omega^t} s_t(\omega^{t-1}) dP^t(\omega^{t-1}) = F^1(l_t^1, k_t^1) + (1-\delta)k_t \tag{23}$$

Proof See the Appendix.

The left-hand side of (23) is the total demand for intermediate goods, and the right-hand side is the total supply. As seen in (4), the intermediate goods are used by the more highly productive agents who draw $\Phi_t(\omega_t) > \phi^*$ to produce capital. Then, Lemmas 3.3 and (4) with the i.i.d. assumption regarding the idiosyncratic productivity shocks yield capital k_{t+1}, as in Proposition 3.2.

Proposition 3.2

$$k_{t+1} = \frac{H(\phi^*)}{1-\mu}\left(F^1(l_t^1, k_t^1) + (1-\delta)k_t\right), \tag{24}$$

where $H(\phi^*) = \int_{\phi^*}^{\eta} \Phi_t(\omega_t) dG(\Phi)$.

Proof See the Appendix.

Define $M(\mu) := H(\phi^*)/(1-\mu)$ in (24). Then, $M(\mu)$ can be considered as the aggregate productivity in the economy. By applying L'Hôpital's rule, one can prove that as $\mu \to 1$, i.e., as the financial market approaches perfection, it follows that $M(\mu) \to \eta$. This outcome means that only the agents who draw the highest productivity become capital producers and the other agents become lenders when the financial market is perfect. In this case, allocative inefficiency regarding the intermediate goods is perfectly corrected, and the highest aggregate productivity is achieved. In contrast, we obtain $M(0) = H(0)$, which is equal to the mean of $\Phi_t(\omega_t)$. When $\mu = 0$, there is no financial market, and no agent can be a lender or a borrower. Instead, all agents become capital producers. The range of variation of $M(\mu)$ depends on the size of the support of the idiosyncratic productivity shocks. Regarding $M(\mu)$, Lemma 3.4 is formally obtained.

Lemma 3.4 *As μ increases from 0 to 1, the aggregate productivity, $M(\mu)$, also increases from $M(0)$ to η, where $M(0)$ is the mean of the idiosyncratic productivity shocks.*

Proof See the Appendix.

Inserting (18) into (24) yields a dynamic equation with respect to capital as follows:

$$k_{t+1} = M(\mu)\left(1 - \delta - \frac{\alpha_L^2 v(p_t)}{\Delta}\right)k_t + \frac{M(\mu)\alpha_K^2}{\Delta}w(p_t). \tag{25}$$

$G(\phi^*) = \mu$ and $\phi^* = r_{t+1}/q_{t+1}$ are used to compute the expected return in Proposition 3.3 below.

Proposition 3.3

$$E[R_{t+1}(\omega_t)|\omega^{t-1}] = q_{t+1}M(\mu). \tag{26}$$

Proof See the Appendix.

Equations (12) and (26) rewrite (7) as follows:

$$\frac{p_{t+1}}{\Lambda p_{t+1}^{\frac{-\theta_1}{\theta_2-\theta_1}} + 1 - \delta} = \beta M(\mu) p_t, \tag{27}$$

which is a dynamic equation with respect to the price of consumption goods.

3.5 Steady State

Assumption 3.2

$$0 < \eta < \frac{1}{1-\delta}$$

Assumption 3.2 implies that $(1-\delta)M(\mu) < 1$ for all $\mu \in [0,1)$. Under Assumption 3.2, $(1-\delta)M(\mu)$ varies from $(1-\delta)H(0)$ to $(1-\delta)\eta$ as μ increases from 0 to 1. In the model with a perfect financial market, the aggregate productivity in capital production is constant. In contrast, the current model allows for the aggregate productivity, $M(\mu)$, to vary from $H(0)$ to η, whose upper limit is $1/(1-\delta)$.

Assumption 3.3 *$\theta_2 > \theta_1$ and $\Delta = \alpha_L^1 \alpha_K^2 - \alpha_L^2 \alpha_K^1 > 0$.*

In Benhabib and Nishimura's model, when the utility function is linear with respect to consumption, indeterminacy of equilibrium always occurs if the intermediate goods are capital intensive from the social perspective, i.e., $\theta_2 > \theta_1$, and labor intensive from the private perspective, i.e., $\Delta > 0$. Therefore, we exclusively examine the case in which $\theta_2 > \theta_1$ and $\Delta > 0$ to investigate whether indeterminacy occurs when the financial market is imperfect.

Under Assumption 3.2, (27) provides the consumption goods price, p^*, in the steady state, as follows:

$$p^* = \left(\frac{\beta \Lambda M(\mu)}{1-(1-\delta)\beta M(\mu)}\right)^{\frac{\theta_2-\theta_1}{\theta_1}}. \tag{28}$$

Furthermore, under Assumption 3.2, (12), (25), and (28) yield the capital stock, k^*, in the steady state as follows:

$$k^* = \frac{\alpha_K^2 \Psi \Lambda^{\frac{1-\theta_1}{\theta_1}} (\beta M(\mu))^{\frac{1}{\theta_1}}}{\left(\alpha_L^2 + \beta\Delta - (\Delta + \alpha_L^2)\beta(1-\delta)M(\mu)\right)(1-(1-\delta)\beta M(\mu))^{\frac{1-\theta_1}{\theta_1}}}. \tag{29}$$

To confirm that the economy produces both intermediate and consumption goods in the steady state, we obtain Lemma 3.5 below.

Lemma 3.5 *Under Assumptions* 3.2 *and* 3.3, *it holds that*

$$\frac{\alpha_K^1 w(p^*)}{\alpha_L^1 v(p^*)} < k^* < \frac{\alpha_K^2 w(p^*)}{\alpha_L^2 v(p^*)}.$$

Proof See the Appendix.

From (18) and (19), Lemma 3.5 implies that the economy imperfectly specializes in production and consistently produces both intermediate and consumption goods in the steady state. By continuity, both intermediate and consumption goods are produced in the neighborhood of the steady state. Throughout the analysis, we exclusively focus on the case in which the economy produces both intermediate and consumption goods.

4 Local Dynamics

From (25) and (27), the dynamical system with respect to k_t and p_t is given by

$$\begin{cases} k_{t+1} = J(k_t, p_t) \\ \dfrac{p_{t+1}}{\Lambda p_{t+1}^{\frac{-\theta_1}{\theta_2-\theta_1}} + 1 - \delta} = \beta M(\mu) p_t, \end{cases} \tag{30}$$

where

$$J(Y_t, p_t) = M(\mu)\left(1 - \delta - \frac{\alpha_L^2 v(p_t)}{\Delta}\right) k_t + \frac{M(\mu)\alpha_K^2}{\Delta} w(p_t).$$

Note that the second equation is expressed by the consumption price only, because we assume that agents' period-wise utility is linear with respect to consumption. The linearization of the dynamical system (30) around the steady state is obtained as

$$\begin{pmatrix} k_{t+1} - k^* \\ p_{t+1} - p^* \end{pmatrix} = \begin{pmatrix} \frac{(\Delta+\alpha_L^2)(1-\delta)\beta M(\mu) - \alpha_L^2}{\beta\Delta} & J_p(Y^*, p^*) \\ 0 & \frac{\theta_2-\theta_1}{\theta_2-\theta_1(1-\delta)\beta M(\mu)} \end{pmatrix} \begin{pmatrix} k_t - k^* \\ p_t - p^* \end{pmatrix}, \tag{31}$$

where $J_p(k, p) := \partial J(k, p)/\partial p$. The eigenvalues, κ_1 and κ_2, of this dynamical system are given by

$$\kappa_1 = \frac{(\Delta + \alpha_L^2)(1 - \delta)\beta M(\mu) - \alpha_L^2}{\beta\Delta} \tag{32}$$

and

$$\kappa_2 = \frac{\theta_2 - \theta_1}{\theta_2 - \theta_1(1-\delta)\beta M(\mu)}. \tag{33}$$

To focus on a typical interesting case, Assumption 4.4 below is imposed in what follows.

Assumption 4.4 $1 < \alpha_L^2/(\beta\Delta) < 1/(1-\beta)$.

The value of κ_1 is crucial for the determination of the dynamic property around the steady state, although it is easily shown that $\kappa_2 \in (0, 1)$, as proven in Lemma A.1 in the Appendix. Figure 2 illustrates the relationship between $M(\mu)$ and κ_1 under Assumption 4.4. In Fig. 2, two critical values of $M(\mu)$ are defined: $M_1 := (\alpha_L^2 - \beta\Delta)/[(\Delta + \alpha_L^2)(1-\delta)\beta]$ and $M_2 := \alpha_L^2/[(\Delta + \alpha_L^2)(1-\delta)\beta]$. We also define μ_1 and μ_2, if any, such that $M(\mu_1) = M_1$ and $M(\mu_2) = M_2$. As noted in Fig. 2, M_1 gives $\kappa_1 = -1$ and M_2 gives $\kappa_1 = 0$. Note also that when $M(\mu) = 1/(1-\delta)$, we have $\kappa_1 = 1 - (1-\beta)\alpha_L^2/(\beta\Delta)$, which is less than 1 and is greater than 0 from Assumption 4.4. The value of κ_1 varies depending on the extent of financial constraints.

Theorem 4.1 *Consider the linearized dynamical system* (31). *Under Assumptions* 2.1, 3.2, 3.3, *and* 4.4, *suppose that the mean of the stochastic productivity shocks,* $M(0)$, *is smaller than* M_1 *and that the maximum,* η, *is greater than* M_2. *Then, if the financial constraint is severe, the steady state is a saddle point, and if the financial constraint is relaxed, the steady state is totally stable. More concretely, the following hold.*

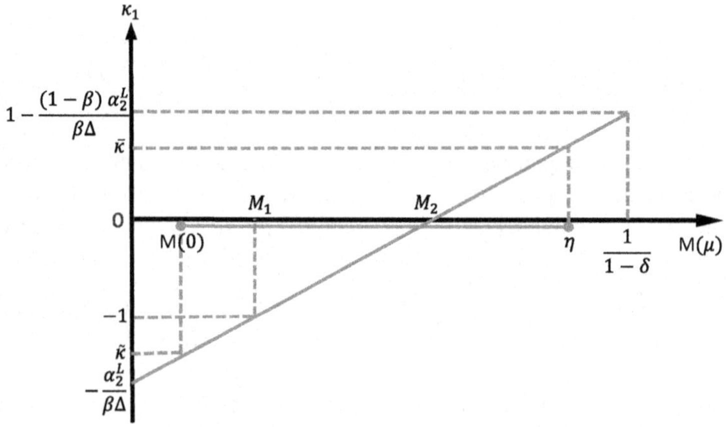

Fig. 2 The relationship between $M(\mu)$ and κ_1. *Notes*: From Assumption 4.4, it follows that $1 - (1-\beta)\alpha_2^L/(\beta\Delta) \in (0, 1)$ and $-\alpha_2^L/(\beta\Delta) \in (-\infty, -1)$ on κ_1 axis

- *If* $\mu \in [0, \mu_1)$*, the steady state is a saddle point.*
- *If* $\mu \in (\mu_1, 1)$*, the steady state is totally stable.*

Proof See the Appendix.

In the dynamical system given by (30), k_t is a state variable that is predetermined at time t, and p_t is a jump variable. In Theorem 4.1 when the steady state is a saddle point, for any given initial capital, k_0, there exists only an initial price of consumption goods that yields a sequence $\{k_t, p_t\}_{t=0}^{\infty}$ in competitive equilibrium. Accordingly, in this case, equilibrium is uniquely determined. In contrast, when the steady state is totally stable, there exists a continuum of initial prices of consumption goods, each one of which produces a sequence $\{k_t, p_t\}_{t=0}^{\infty}$ in competitive equilibrium. In this case, equilibrium is indeterminate and multiple equilibria occur.

It is widely known that indeterminacy of equilibrium causes self-fulfilling sunspot business fluctuations (Shell [22]; Azariadis [4]; Cass and Shell [12]; Woodford [23]). Extrinsic random variables that may have impacts on agents' expectations without directly affecting economic fundamentals are called sunspots. If the resource allocation in equilibrium depends on the realization of a sunspot random variable, the equilibrium is called a sunspot equilibrium. When indeterminacy arises, extrinsic uncertainty randomizes multiple equilibria. Then, sunspot business fluctuations can occur as a rational expectations equilibrium. Note from Theorem 4.1 that when the financial constraint is severe, no self-fulfilling sunspot business fluctuations occur, whereas when the financial constraint is relaxed, self-fulfilling sunspot business fluctuations can occur.

5 Concluding Remarks

A two-sector dynamic general equilibrium model in which agents face financial constraints and the production sectors exhibit externalities is presented. Whether production externalities cause indeterminacy of equilibria depends on the extent of financial constraints and the size of the support of the idiosyncratic productivity shocks. Under the moderate parameter conditions for a labor income share and sector-specific production externalities, if financial constraints are severe, equilibrium is unique. However, as financial constraints are relaxed, equilibrium is indeterminate, and thus, self-fulfilling sunspot business fluctuations can occur.

Recently, it has often been asserted that financial innovations that relax financial constraints destabilize economies. The outcomes from our analysis are consistent with this assertion. In our model, the relaxation of financial constraints magnifies the destabilization effect of production externalities.

Acknowledgements The authors would like to express thanks to Toru Maruyama, who is an editor, for his comments. All remaining errors, if any, are ours. This work is financially supported by the Japan Society for the Promotion of Science, Grants-in-Aid for Scientific Research (Nos. 15H05729, 16H02026, 16H03598, 16K03685).

Appendix

Proof of Proposition 3.1

It follows from Lemma 3.1 that

$$\int_{\Xi_t}\int_{\Omega^t} s_t(\omega^{t-1})dP^t(\omega^{t-1})dP(\omega_t) - \frac{\mu}{1-\mu}\int_{\Omega\setminus\Xi_t}\int_{\Omega^t} s_t(\omega^{t-1})dP^t(\omega^{t-1})dP(\omega_t) = 0. \tag{34}$$

where $\Xi_t = \{\omega_t \in \Omega | \Phi_t(\omega_t) \le \phi_t\}$. (34) can be rewritten as

$$\int_{\Omega^t} s_t(\omega^{t-1})dP^t(\omega^{t-1})\int_0^{\phi_t} dG(\Phi) - \frac{\mu}{1-\mu}\int_{\Omega^t} s_t(\omega^{t-1})dP^t(\omega^{t-1})\int_{\phi_t}^{\eta} dG(\Phi) = 0,$$

which is computed as

$$\int_{\Omega^t} s_t(\omega^{t-1})dP^t(\omega^{t-1})\left[G(\phi_t) - \frac{\mu}{1-\mu}(1-G(\phi_t))\right] = 0.$$

Solving the last part, we obtain $G(\phi_t) = \mu$. □

Proof of Lemma 3.2

Because $R_t(\omega_{t-1}) = \max\{r_t, (q_t\Phi_{t-1}(\omega_{t-1}) - r_t\mu)/(1-\mu)\}$, it follows that

$$\int_{\Omega^t} R_t(\omega_{t-1})s_{t-1}(\omega^{t-2})dP^t(\omega^{t-1}) = \int_{\Omega^t} \max\left\{r_t, \frac{q_t\Phi_{t-1}(\omega_{t-1}) - r_t\mu}{1-\mu}\right\} \times s_{t-1}(\omega^{t-2})dP^t(\omega^{t-1}) =: I_t$$

Define $\Xi_{t-1} = \{\omega_{t-1} \in \Omega | \Phi_{t-1}(\omega_{t-1}) \le \phi_{t-1}\}$, as in the proof of Proposition 3.1. Because $\phi_{t-1} = r_t/q_t$, I_t can be computed as follows:

$$I_t = \int_{\Omega^{t-1}\times\Xi_{t-1}} r_t s_{t-1}(\omega^{t-2})dP^t(\omega^{t-1})$$

$$
\begin{aligned}
&+\int_{\Omega^{t-1}\times(\Omega\backslash\Xi_{t-1})}\frac{q_t\Phi_{t-1}(\omega_{t-1})-r_t\mu}{1-\mu}s_{t-1}(\omega^{t-2})dP^t(\omega^{t-1})\\
&=\int_{\Omega^{t-1}\times(\Omega\backslash\Xi_{t-1})}\frac{q_t\Phi_{t-1}(\omega_{t-1})}{1-\mu}s_{t-1}(\omega^{t-2})dP^t(\omega^{t-1}),
\end{aligned}
\tag{35}
$$

where the second equality of (35) is obtained from Lemma 3.1. Agents who draw $\Phi_{t-1}(\omega_{t-1}) > \phi_{t-1}$ invest $x_{t-1}(\omega^{t-1}) = s_{t-1}(\omega^{t-2})/(1-\mu)$, and thus, (35) becomes

$$
\begin{aligned}
I_t &= \int_{\Omega^{t-1}\times(\Omega\backslash\Xi_{t-1})}\frac{q_t\Phi_{t-1}(\omega_{t-1})}{1-\mu}(1-\mu)x_{t-1}(\omega^{t-1})dP^t(\omega^{t-1})\\
&= q_t\int_{\Omega^{t-1}\times(\Omega\backslash\Xi_{t-1})}\Phi_{t-1}x_{t-1}(\omega^{t-1})dP^t(\omega^{t-1}).
\end{aligned}
\tag{36}
$$

Because $k_t = \int_{\Omega^{t-1}\times(\Omega\backslash\Xi_{t-1})}\Phi_{t-1}x_{t-1}(\omega^{t-1})dP^t(\omega^{t-1})$, (36) becomes

$$
I_t = q_t k_t. \ \square
$$

Proof of Lemma 3.3

By using Lemma 3.2 and because $\int_{\Omega^t}\pi_t dP^t(\omega^{t-1}) = \Pi_t$, the aggregation of (6) across all agents is obtained as follows:

$$
\begin{aligned}
\int_{\Omega^t}s_t(\omega^{t-1})dP^t(\omega^{t-1}) &= \int_{\Omega^t}[R_t(\omega_{t-1})s_{t-1}(\omega^{t-2})+w_t+\pi_t\\
&\quad - p_t c_t(\omega^{t-1})]dP^t(\omega^{t-1})\\
&= q_t k_t + w_t + \Pi_t - p_t C_t.
\end{aligned}
\tag{37}
$$

From (8), we have $F^1(l_t^1,k_t^1)+p_tF^2(l_t^2,k_t^2)+(1-\delta)k_t = q_tk_t+w_t+\Pi_t$. Additionally, the consumption goods market-clearing condition leads to $p_tF^2(l_t^2,k_t^2) = p_tC_t$. Therefore, (37) is transformed into

$$
\int_{\Omega^t}s_t(\omega^{t-1})dP^t(\omega^{t-1}) = F^1(l_t^1,k_t^1)+(1-\delta)k_t. \ \square \tag{38}
$$

Proof of Proposition 3.2

Because k_{t+1} is produced by capital producers who draw an individual-specific productivity, $\Phi_t(\omega_t)$, that is greater than ϕ_t, Lemma 3.3 and the i.i.d. assumption compute k_{t+1} as follows:

$$\begin{aligned} k_{t+1} &= \int_{\Omega^t \times (\Omega \setminus \Xi_t)} \Phi_t(\omega_t) x_t(\omega^t) dP^{t+1}(\omega^t) \\ &= \int_{\Omega \setminus \Xi_t} \int_{\Omega^t} \Phi_t(\omega_t) \frac{s_t(\omega^{t-1})}{1-\mu} dP^t(\omega^{t-1}) dP(\omega_t) \\ &= \int_{\phi_t}^{\eta} \frac{\Phi_t(\omega_t)}{1-\mu} dG(\Phi) \int_{\Omega^t} s_t(\omega^{t-1}) dP(\omega^{t-1}) \\ &= \frac{H(\phi^*)}{1-\mu} (F^1(l_t^1, k_t^1) + (1-\delta)k_t), \end{aligned}$$

where $H(\phi^*) = \int_{\phi^*}^{\eta} \Phi_t(\omega_t) dG(\Phi)$ because $\phi_t = \phi^*$ in equilibrium. □.

Proof of Lemma 3.4

Obviously, $M(\mu)$ is continuous in $[0, 1)$. The inverse function theorem implies

$$\begin{aligned} \frac{\partial M(\mu)}{\partial \mu} = \frac{\partial}{\partial \mu}\left(\frac{H(\phi^*)}{1-\mu}\right) &= \frac{-(1-\mu)\phi^* G'(\phi^*)(\partial \phi^*/\partial \mu) + H(\phi^*)}{(1-\mu)^2} \\ &= \frac{\int_{\phi^*}^{h} \Phi_t(\omega_t) dG(\Phi) - \phi^*(1-G(\phi^*))}{(1-\mu)^2} > 0. \end{aligned}$$

Therefore, $M(\mu)$ is an increasing function with respect to μ in $[0, 1)$. It is straightforward to verify that $M(0) = H(0)$ is the mean of the idiosyncratic productivity shocks. By applying L'Hôpital's rule, we obtain $\lim_{\mu \to 1} M(\mu) = \lim_{\mu \to 1} G^{-1}(\mu) G^{-1'}(\mu) G'(\phi^*) = \eta$. For the last equality, we have used the inverse function theorem again. □

Proof of Proposition 3.3

Because $\phi_t = r_{t+1}/q(p_{t+1})$ and $\phi_t = \phi^*$, it follows that

$$\begin{aligned}E[R_{t+1}(\omega_t)|\omega^{t-1}] &= E\left[\max\left\{r_{t+1}, \frac{q_{t+1}\Phi_t(\omega_t) - r_{t+1}\mu}{1-\mu}\right\}\middle|\omega^{t-1}\right]\\ &= q_{t+1}E\left[\max\left\{\phi_t, \frac{\Phi_t(\omega_t) - \phi_t\mu}{1-\mu}\right\}\middle|\omega^{t-1}\right]\\ &= q_{t+1}\left[\int_0^{\phi^*}\phi^* dG(\Phi) + \int_{\phi^*}^{\eta}\frac{\Phi_t(\omega_t) - \phi^*\mu}{1-\mu}dG(\Phi)\right]\\ &= q_{t+1}\left[\phi^* G(\phi^*) - \frac{\phi^*\mu}{1-\mu}(1-G(\phi^*)) + \frac{H(\phi^*)}{1-\mu}\right]\\ &= q_{t+1}M(\mu).\end{aligned}$$

To derive the last equality, Proposition 3.1 is applied. □

Proof of Lemma 3.5

From Assumption 3.2, it follows that $(1-\delta)M(\mu) < 1$. Then, under Assumption 3.3, from (12), (28), and (29), it follows that $sign\{k^* - \alpha_K^1 w(p^*)/(\alpha_L^1 v(p^*))\} = sign\{1 - (1-\delta)\beta M(\mu) - \alpha_K^1\beta(1-(1-\delta)M(\mu))\}$. Because $1-(1-\delta)\beta M(\mu) - \alpha_K^1\beta(1-(1-\delta)M(\mu)) > (1-\alpha_K^1\beta)(1-(1-\delta)M(\mu)) > 0$, it follows that $sign\{k^* - \alpha_K^1 w(p^*)/(\alpha_L^1 v(p^*))\} > 0$. Additionally, it follows that $sign\{\alpha_K^2 w(p^*)/(\alpha_L^2 v(p^*)) - k^*\} = sign\{1-(1-\delta)M(\mu)\} > 0$. □

Proof of Theorem 4.1

To prove Theorem 4.1, two lemmata are prepared.

Lemma A.1 *Under Assumptions* 2.1, 3.2, *and* 3.3, *it holds that* $0 < \kappa_2 < 1$.

Proof It follows from Assumption 3.2 that $0 < (1-\delta)\beta M(\mu) < 1$, which leads to $(\theta_2 - \theta_1)/(\theta_2 - \theta_1(1-\delta)\beta M(\mu)) < 1$ from Assumptions 2.1, 3.2, and 3.3. Additionally, it follows from Assumption 3.3 that $(\theta_2-\theta_1)/(\theta_2-\theta_1(1-\delta)\beta M(\mu)) > 0$

□

Lemma A.2 *Under Assumptions* 2.1, 3.2, 3.3, *and* 4.4, *suppose that the mean of the stochastic productivity shocks,* $M(0)$, *is smaller than* M_1 *and that the maximum,* η, *is greater than* M_2. *Then, as the value of* μ *increases from* 0 *to* 1, *the value of* κ_1 *increases in the ranges, as in the following.*

- *As* μ *increases in* $[0, \mu_1)$, κ_1 *increases in* $[\tilde{\kappa}, -1)$.
- *As* μ *increases in* $[\mu_1, \mu_2)$, κ_1 *increases in* $[-1, 0)$.
- *As* μ *increases in* $[\mu_2, 1)$, κ_1 *increases in* $[0, \bar{\kappa})$,

where $\tilde{\kappa} := [(\Delta + \alpha_L^2)(1-\delta)\beta M(0) - \alpha_L^2]/(\beta\Delta) \in (-\infty, -1)$ *and* $\bar{\kappa} < 1 - (1 - \beta)\alpha_L^2/(\beta\Delta) \in (0, 1)$, *which is given when* $M(\mu) = \eta$.

Proof $M(\mu)$ is an increasing function with respect to μ. Then, Fig. 2 and (32) prove the claims. □

Proof of Theorem 4.1 From Lemma A.1, we have $|\kappa_2| < 1$. From Lemma A.2, if $\mu \in [0, \mu_1)$, $|\kappa_1| > 1$, and if $\mu \in (\mu_1, 1)$, $|\kappa_1| < 1$. Therefore, the steady state is a saddle point if $\mu \in [0, \mu_1)$, and the steady state is totally stable if $\mu \in (\mu_1, 1)$. □

References

1. Aghion P, Banerjee A (2005) Volatility and growth. Oxford University Press, New York
2. Aghion P, Banerjee A, Piketty T (1999) Dualism and macroeconomic volatility. Q J Econ 114(4):1359–1397
3. Aghion P, Howitt P, Mayer-Foulkes D (2005) The effect of financial development on convergence: theory and evidence. Q J Econ 120(1):173–222
4. Azariadis C (1981) Self-fulfilling prophecies. J Econ Theory 25(3):380–396
5. Barinci J-P, Chéron A (2001) Sunspots and the business cycle in a finance constrained economy. J Econ Theory 97(1):30–49
6. Benhabib J, Farmer REA (1994) Indeterminacy and increasing returns. J Econ Theory 63(1):19–41
7. Benhabib J, Farmer REA (1996) Indeterminacy and sector-specific externalities. J Monet Econ 37(3):421–443
8. Benhabib J, Nishimura K (1998) Indeterminacy and sunspots with constant returns. J Econ Theory 81(1):58–96
9. Benhabib J, Wang P (2013) Financial constraints, endogenous markups, and self-fulfilling equilibria. J Monet Econ 60:789–805
10. Benhabib J, Meng Q, Nishimura K (2000) Indeterminacy under constant returns to scale in multisector economies. Econometrica 68(6):1541–1548
11. Boldrin M, Rustichini A (1994) Growth and indeterminacy in dynamic models with externalities. Econometrica 62(2):323–342
12. Cass D, Shell K (1983) Do sunspots matter? J Polit Econ 91(2):193–227
13. Cordoba J-C, Ripoll CM (2004) Collateral constraints in a monetary economy. J Eur Econ Assoc 2(6):1172–1205
14. Cordoba J-C, Ripoll M (2004) Credit cycles redux. Int Econ Rev 45(4):1011–1046
15. Dufourt F, Nishimura K, Venditti A (2015) Indeterminacy and sunspots in two-sector RBC models with generalized no-income-effect preferences. J Econ Theory 157:1056–1080
16. Kiyotaki N, Moore J (1997) Credit cycles. J Polit Econ 105(2):211–248
17. Kunieda T, Shibata A (2016) Asset bubbles, economic growth, and a self-fulfilling financial crisis. J Monet Econ 82:70–84
18. Liu Z, Wang P (2014) Credit constraints and self-fulfilling business cycles. Am Econ J Macroecon 6(1):32–69
19. Nishimura K, Shimomura K (2002) Trade and indeterminacy in a dynamic general equilibrium model. J Econ Theory 105(1):244–260
20. Nishimura K, Venditti A (2004) Indeterminacy and the role of factor substitutability. Macroecon Dyn 8(04):436–465
21. Nishimura K, Venditti A (2007) Indeterminacy in discrete-time infinite-horizon models with non-linear utility and endogenous labor. J Math Econ 43(3–4):446–476

22. Shell K (1977) Monnaie et allocation intertemporelle. Mimeograph, CNRS Seminaire Roy-Malinvaud (Paris), November 21, 1977
23. Woodford M (1986) Stationary sunspot equilibria: the case of small fluctuations around a deterministic steady state. Mimeo, University of Chicago
24. Woodford M (1986) Stationary sunspot equilibria in a finance constrained economy. J Econ Theory 40(1):128–137

Index

S. Kusuoka, T. Maruyama (eds.), *Advances in Mathematical Economics*, Advances in Mathematical Economics 22, https://doi.org/10.1007/978-981-13-0605-1

Printed in the United States
By Bookmasters